History of Life
Course ePack

2012 edition

David A. DeWitt, Ph.D.

"When I consider your heavens, the work of your fingers,
the moon and the stars which you have set in place,
what is man that you are mindful of him, the son of man that you care for him?"
Psalm 8:3-4

History of Life Course ePack, 2012 edition
Copyright © 2012 by David A. DeWitt, Ph.D.

Scripture taken from the HOLY BIBLE, NEW INTERNATIONAL VERSION. Copyright © 1973, 1978, 1984 International Bible Society. Used by permission of Zondervan Bible Publishers.

All rights reserved. No part of this publication may be reproduced or transmitted in any form or by any means, electronic or mechanical, including but not limited to photocopying, recording, or any information storage and retrieval system, without written permission from the author and the publisher.

Requests for permission should be sent to:
Creation Curriculum LLC
P.O. Box 4662
Lynchburg, VA 24502

Printed in USA

ISBN: 978-0-9796323-4-1

PREFACE

Welcome to Creation Studies 290!

The creation/evolution debate is not a new one but has certainly increased in intensity over the last few years. In 1 Peter 3:15, Peter admonishes Christians to "always be prepared to give an answer." Evolution is an obstacle for many people and prevents them from trusting the Bible. They think that the accounts of creation and Noah's Flood are myths and therefore the Bible is just a book of fairy tales rather than the Word of God. Thus, it is important for all believers to be prepared to give answers to questions and challenges that are raised against Genesis.

This course on origins has been specially designed to equip students to defend their faith and give answers to common questions that people have about creation and evolution. Since most of the evidence and arguments about origins and evolution is scientific, the majority of the material in the course is about science. However, it is science presented from a Biblical creation perspective. The emphasis on scientific evidence does not imply that this is an esoteric subject suitable only for those trained in the sciences. All Christians should have a basic understanding of the scientific arguments and evidence regarding origins.

This study guide notebook that accompanies the course will help you with your note taking and provide you with most of the text from the lecture slides. Throughout the book are blanks for you to fill in. This is to help you follow along and key in to important concepts. However, you should also take your own notes along the margins and in the blank spaces. Information will be given in lecture that does *not* appear in the notebook—it is information that you should write down yourself! Remember, the content of the notebook is not the only source of material for the exams. Nonetheless, the notebook will be helpful for you as a central location for notes and as a guide to the most important material.

This ePack includes the study guide notebook, online access to a number of videos and lectures as well as online access to the textbook: Unraveling the Origins Controversy. Please note that your exam questions will be coming from all of these sources.

I trust that you will find this class informative and one that will encourage you in your Christian faith. It is my prayer that it will also strengthen your Biblical worldview and will provide answers that you can share with others. Use this information to promote the Gospel of Jesus Christ and lead others to saving faith in Him.

Amen.

David A. DeWitt, Ph.D.

Director, Center for Creation Studies

Liberty University

TABLE OF CONTENTS

Lesson 1

Introduction

APOLOGETICS

- Branch of theology which deals with the ________________ and _________________

- Does not mean "to say you're sorry"

- Derived from _______________

1 Peter 3:15

"…always be prepared to give an απολογια (_______) to everyone who asks you to give the reason for the hope that you have, *but do this with* _____________________." (NIV)

Acts 25:8

"Then Paul made his απολογια…"

As Christians, we must be prepared to give an ________, a defense, an __________, for the reason for our hope in Christ.

Proverbs 19:2

"It is not good to have ______ without knowledge nor to be hasty and miss the way" (NIV)

2 Peter 1:5

"For this very reason, make every effort to add to your faith goodness; and to goodness, _____________…" (NIV)

Course Purpose

To equip you with knowledge, arguments and evidence relating to the creation/evolution debate.

Provide the facts you need to give an answer to anyone who asks why you believe that *God created* the universe.

1 Peter 3:15

- "…always be prepared to give an απολογια (______) to everyone who asks you to give the reason for the hope that you have, *but do this with*______________________."

Gentleness and Respect?

- Psalm 14:1 "The _______ has said in his heart, 'There is no God.'"
- Galileo
- Don't win the debate but lose the war

Theory of Evolution

- "Process by which ________________________. The idea that all ____."

  -Tobin & Dusheck 1998
- Offered to explain the origin of all life, the universe, and everything.
- Differences arise through __________ and _________________

Evolutionary Tree

Suggests ALL living things share a ________________

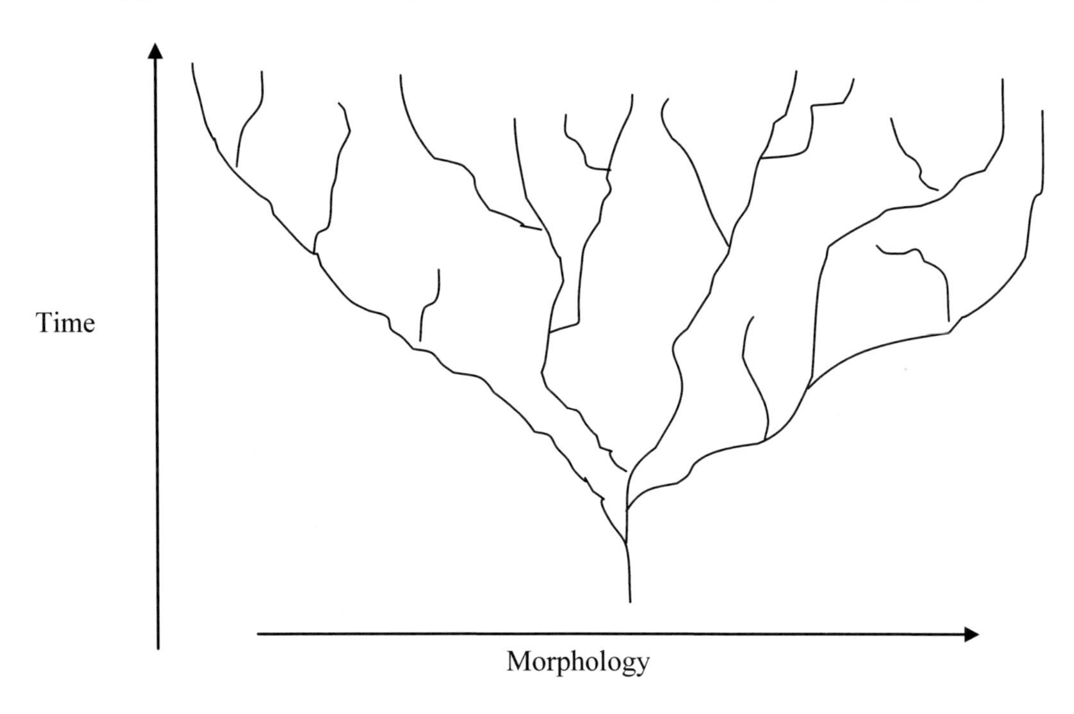

The "Creation Orchard"

- Different kinds of organisms were created with a "_______ of variability".
- This allows for _____, __________ and ________ since creation

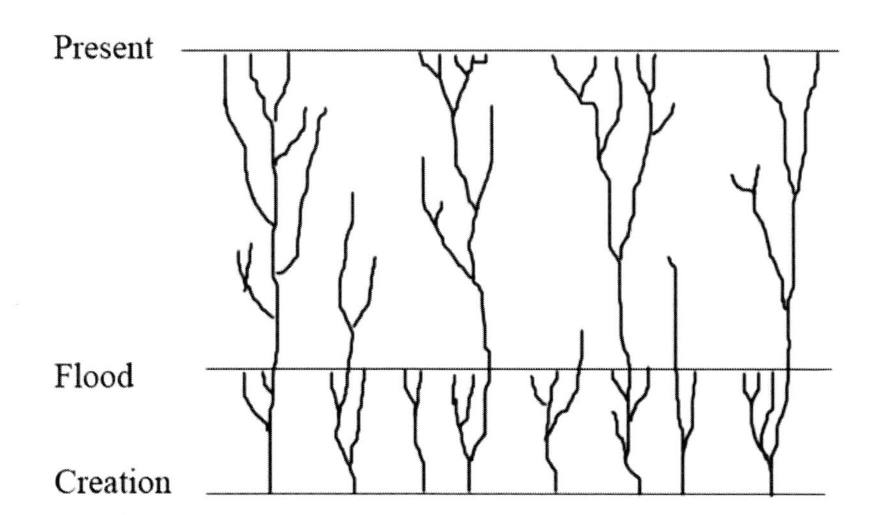

Lesson 2

Significance of Creation

Outcomes of an Evolution View

- Racism
- Forced sterilization (Eugenics)
- __________

Declaration of Independence

- "We hold these truths to be self evident: That all men are created equal, that they are endowed by their Creator with certain unalienable rights, that among these are life, liberty, and the pursuit of happiness."

Declaration of Independence?

- We hold this fact to be proven, that all men have evolved and therefore are not equal because some are more fit than others. Further, since there is no 'Creator' there are no unalienable rights except the survival of the fittest.

____

- Textbook used by John _________ 1925

Man then began to cultivate the fields, and to have a fixed place of abode other than a cave. The beginnings of civilization were long ago, but even to-day the earth is not entirely civilized.

The Races of Man. — At the present time there exist upon the earth five races or varieties of man, each very different from the other in instincts, social customs, and, to an extent, in structure. These are the Ethiopian or negro type, originating in Africa; the Malay or brown race, from the islands of the Pacific; the American Indian; the Mongolian or yellow race, including the natives of China, Japan, and the Eskimos; and finally, the highest type of all, the Caucasians, represented by the civilized white inhabitants of Europe and America.

Hunter's Civic Biology

“Biological arguments for racism may have been common before 1850, but they ___________ by orders of magnitude following the acceptance of evolution theory.”

Stephen J. Gould Ontogeny and Phylogeny, Belknap-Harvard Press 1977 p 127-128

“The Darwinian revolution and the works of its chief German spokesman and most eminent scientist, Professor Haeckel, gave the racists something that they were confident was powerful verification of their race beliefs. The support of the _________ __________ resulted in racist thought having a much wider circulation than otherwise possible, and enormous satisfaction ‘that one’s prejudices were actually expressions of ______________’”

Schleunes, K., The Twisted Road to Auschwitz, University of Illinois Press, Urbana IL 1970.

No Basis for Racism

t “From _________ he made every nation of men, that they should inhabit the whole earth; and he determined the times set for them and the exact places where they should live” Acts 17:26

t “‘I now realize how true it is that God does not show favoritism but accepts men from every ______ who fear him and do what is right.’” Peter, Acts 10:35-36

t In Scripture there is only one race: The _________ Race.

t “Adam named his wife Eve because she would become the mother of ___________.” Genesis 3:20

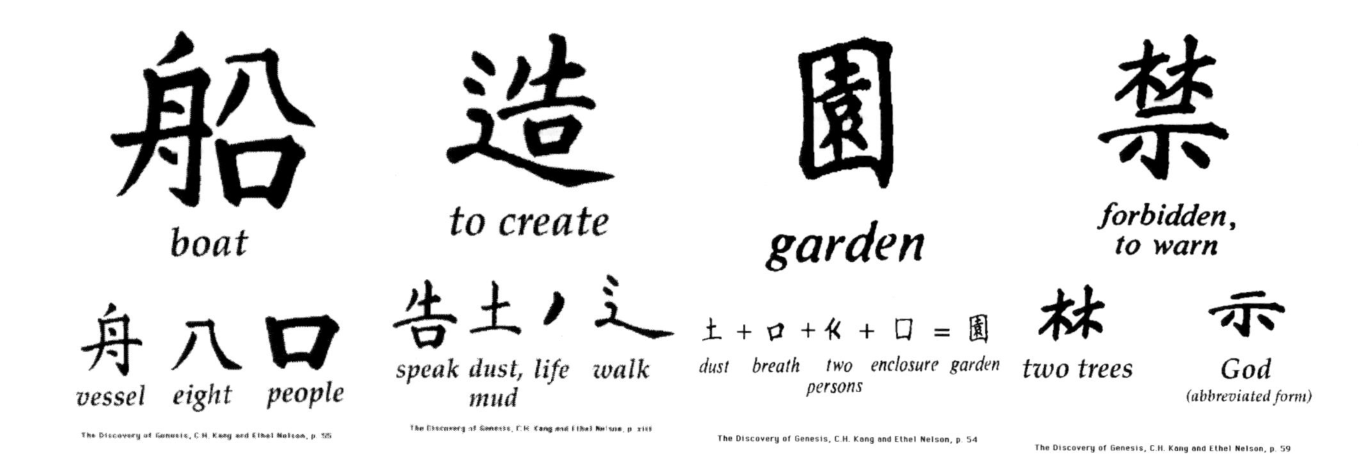

Eugenics (Forced Sterilization)

More than 40,000 people classed as insane or “feebleminded” were ______________ by law in 30 states by 1944.

Another 22,000 were __________ from then until 1963.

Similar laws in Sweden, France, and Australia.

Decision in Buck v. Bell

cannot be justified upon the existing grounds. The judgment finds the facts that have been recited and that Carrie Buck "is the probable potential parent of socially inadequate offspring, likewise afflicted, that she may be sexually sterilized without detriment to her general health and that her welfare and that of society will be promoted by her sterilization," and thereupon makes the order. In view of the general declarations of the legislature and the specific findings of the Court obviously we cannot say as matter of law that the grounds do not exist, and if they exist they justify the result. We have seen more than once that the public welfare may call upon the best citizens for their lives. It would be strange if it could not call upon those who already sap the strength of the State for these lesser sacrifices, often not felt to be such by those concerned, in order to prevent our being swamped with incompetence. It is better for all the world, if instead of waiting to execute degenerate offspring for crime, or to let them starve for their imbecility, society can prevent those who are manifestly unfit from continuing their kind. The principle that sustains compulsory vaccination is broad enough to cover cutting the Fallopian tubes. *Jacobson* v. *Massachusetts*, 197 U. S. 11. Three generations of imbeciles are enough.

"No more children should be born when the parents, though healthy themselves, find that their children are physically or mentally defective. No matter how much they desire children, no man and woman have a right to bring into the world those who are to suffer from mental or physical affliction. It condemns the child to a life of misery and places upon the community the burden of caring for it, probably for its defective descendants for many generations..."

--_______ ___________

Woman and the New Race, 1920

Eugenics now carried out by ___________

Conclusions of Evolution by Natural Selection

Since all living things have a common ancestor no species is "better" than another

Since evolution occurs by ________, it is an "accident" that humans evolved at all.

In *natural* selection the "___" survive and the "_____" die off. To prevent this process is to circumvent evolution.

What then is the value of a human life?

"The preciousness and dignity of the individual person is a prime value."

This is a ________ doctrine.

But...

If we begin in insignificance and end with insignificance, then how can we have significance in between?

Does God value man?

t "How much more valuable is a man than a _____!" Matthew 12:12

t “And how much more valuable you are than _____!” Luke 12:24

t “When I consider your heavens, the work of your fingers…what is man that you are mindful of him?” Psalm 8:3-4

What gives human life value?

t Created “in the ____________”

t Each and every person is someone for whom Christ died

t This is _________ value and worth not based on ability, age, characteristics, or quality of life but ____________

t Therefore every human life is ______.

Can you prove that creation is true?

t Both the evolution and creation views depend on __________.

t What type of ________ are you willing to accept and how much do you need?

Evidence for the existence of God

1. Order, harmony and __________ in the universe and life
2. __________: an intrinsic sense of right and wrong.

These may tell us that there is a God, but doesn’t tell us who that God is or what He requires. For that we need __________.

Creation is a starting point to faith

t Paul in _______

t “Men of Athens, I see that in every way you are very religious…. The God who made the world and everything in it is the LORD of heaven and earth and does not live in temples built by human hands” Acts 17:22, 24

Why believe Genesis?

t Jesus quotes from _______ 1 and 2 in Matthew 19:4-5

t Jesus and Peter both mention ____ as factual history

t Paul talks about ____ (Romans 5) and ___ (1 Timothy 2:13)

Theistic Evolution

Belief that God "used evolution to create"

A "compromise" to blend faith and science

Inconsistencies in order of creation

Inconsistent with God's character of love

Biggest Problem

- Death before sin

"As for God, his way is perfect; the word of the LORD is Flawless. He is a shield for all who take refuge in Him." Psalm 18:30

Lesson 3

Limitations of Science

"The Emperor has no clothes!"

The first to present his case seems _____ till another comes forward and _______________.

– Proverbs 18:17

Can you prove that creation is true?

t Both the evolution and creation views depend on ___________.
t What type of ________ are you willing to accept and how much do you need?

Romans 1:20

"For since the creation of the world God's invisible qualities--his eternal power and divine nature--have been clearly seen, being understood from what has been made, so that men are without excuse."

GODISNOWHERE

____

t One supposes it to be true
t A ____ taken for granted
t Assumptions/presuppositions inform ____ you will look at and interpret data
t They exist ______ investigation begins

Assumptions Are Crucial

"Using our empirical rate to calibrate the mtDNA molecular clock would result in an age of the mtDNA MRCA of only ______ y.a., clearly incompatible with the known age of modern humans."
Nature Genetics 15:365

Evolution ______

Piltdown man
Haeckel's Embryos
Archaeoraptor
Figure by Ernst Haeckel

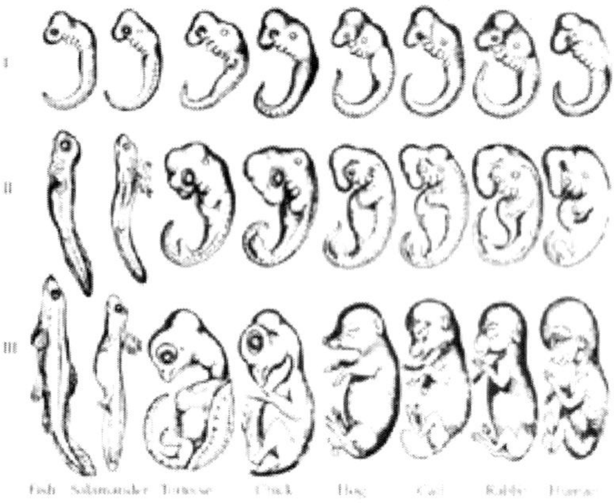

How do we look at the world?

- ♦ Creation assumptions
  - God exists and created all things
  - He told us how he made everything; by His Word _________
- ♦ Evolution assumptions
  - There is no God
  - only ___________ and processes can be used to explain origins

Define "________"

t Pursuit of and possession of knowledge, a body of knowledge

t A systematic investigation into ________________.

Distinguishing Characteristics

t ____________

– examine, measure phenomena

t _________________

– manipulation of phenomena to determine effects (change variables, use control)

t _______________

– results reproducible in time and space
– history isn't repeatable, not empirical

Scientific method (Generally)

- 1. State the ________
- 2. Develop hypothesis
- 3. ____ hypothesis, repeat (experiment)
- 4. ________ and draw conclusions
- 5. Accept or modify hypothesis

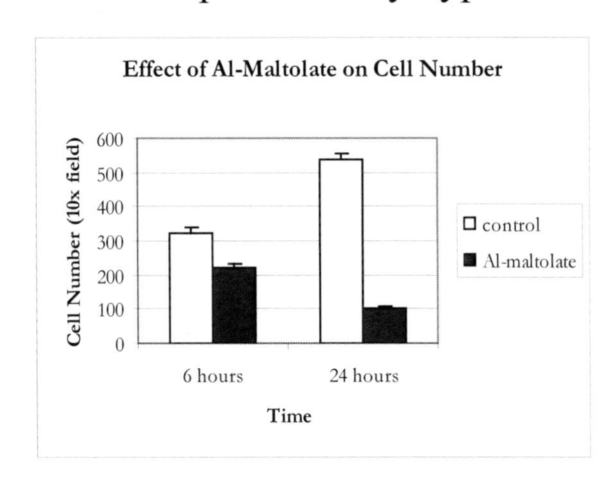

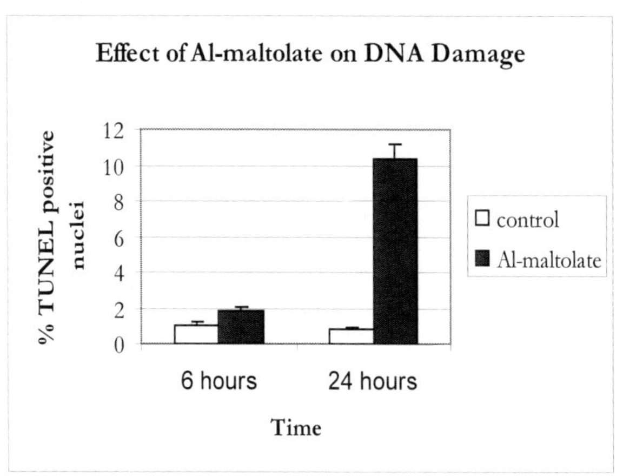

A good hypothesis is ___________

- Scientific method cannot be used to "_____" historical events
- Historical events are "proven" by ____________________

Historical Sciences

- Build models and/or develop hypotheses
- Collect data
- Does data support one or more hypotheses?
- Choose best ___________ among available options

- "Multiple competing ____________"

Types of models

- Substitute
  - replacement when original cannot be used or observed
- Framework
  - complex system used to define, describe, and interpret data

Models

- Must be used with caution!
- The closer the model to the original, the better the model
- Models can be used to make ___________s, we can use the _________________

Mutually _________ Models

- Biblical Creation (literal)
  - Young-Earth Creation
  - Historical-Grammatical Interpretation
- Evolutionary Origin

Creation	Evolution
_____ supernatural acts of God	Natural laws & processes only
_____ age for earth/universe	Very ___ earth and universe
_________ as primary source of truth	Theories subject to ______
Scripture is _______	Scientific/Historic observations only source of truth
Human philosophy second	

Mutually Exclusive

- Cannot both be simultaneously _____
- You cannot be a married bachelor
- You cannot be two different ages

- **Only one primary source of _____**

Creation	Evolution	Models
supernatural	natural	
externally directed	self-contained	
_________	_____________	
maintain/decrease complexity	increase complexity	
___________	_____________	

Predictions: Appearance of life

- Creation: life only from life
- Evolution: life from non-life

Predictions: Fossil Record

Creation:
systematic _____
Distinct ancestors for each kind

Evolution:
transitional _______
Common ancestors

Creation
Time
Morphology

Time
Morphology

- It is impossible to _____ scientifically that any concept of origins is true.

- Biblical creation cannot be proven (or dis-proven) using the scientific method, but must be accepted by _____.

- __________ origin cannot be proven (or dis-proven) using the scientific method, but must be accepted by _____.

Evolution is built on naturalistic _____________

- t “Even if all the data point to an intelligent designer, such an hypothesis is excluded from science because it is not naturalistic.”
  – Dr. Scott Todd, Kansas State University
  correspondence to *Nature* **401**(6752):423
- t
- t Even if scientists could create life from non-life, this would not prove scientifically that such a process occurred __________.

Evolution cannot be _______

- ♦ "Our theory of evolution has become . . .one which cannot be refuted by any possible observation. Every conceivable observation can be fitted into it. It is thus `outside of empirical science' but not necessarily false. No one can think of ways in which to test it. Ideas, either without basis or based on a few laboratory experiments carried out in extremely simplified systems have attained currency far beyond their validly. They have become part of an evolutionary dogma accepted by most of us as part of our training."—**P. Erlich and *L.C. Birch, "Evolutionary History and Population Biology," Nature, Vol. 214, April 22, 1967, p. 352*

Creation is built on _______

- t Now _____ is being sure of what we hope for and certain of what we do not see.”
- t “By______ we understand that the universe was formed at God’s command, so that what is seen was not made out of what was visible.”

  -Hebrews 11:1 and 4 *NIV*

Lecture 4

Icons of Evolution

Notes

Notes

Lessons 5 & 6

Science & Scripture

Creation according to Scripture

- The Historicity of the Genesis Record
- Who wrote Genesis?
  - ___: Genesis 1 to 2
  - ____ Genesis 2 to 5
  - _____ & Sons Genesis 5-11
  - _________ & Sons Genesis 11-50
  - _______: "this is the account of…"

_____ may have compiled and edited these

How was history transmitted?

- Word of mouth through many generations?+
- Not quite…
- ____ was alive with Noah's father (______) and grandfather (__________)
- ____ lived at the same time as Abram
- Therefore Adam-Abram through _ men

The Work of the Six days

Day 1

- _________ and earth, _______ and ________

Day 2

- Firmament, (sky).

Day 3

- _____ and _______

Day 4

- Celestial bodies, _____, ____ and _______

Day 5

- ____ and ______

Day 6

- Land animals
- Man and Woman

Day 7

- God ______ on the seventh day and blessed it.

_______ was involved in Creation

- "In the beginning was the Word, and the Word was with God, and the Word was God. He was with God in the beginning. Through him all things were made; without him nothing was made that has been made."
  -John 1:1-3 (NIV)

- "He is the image of the invisible God, the firstborn over all creation. For by him all things were created: things in heaven and on earth, visible and invisible, whether thrones or powers or rulers or authorities; all things were created by him and for him. He is before all things and in him all things hold together. -Colossians 1:15-16

When was the Fall?

- Must have been _________
- Had to occur before Adam and Eve __________________
- No _____ before the Fall
- God promised a ________

Physical death allows man to be _________
If Adam ate from the tree of life as a sinful being, he would live _______________ from God.

Sequence in Evolution

- ♦ Big Bang, origin of matter
- ♦ Solar system, sun & planets
- ♦ Origin of life in _______
- ♦ __________, unicellular organisms, then _____
- ♦ _________, then land animals (reptiles)
- ♦ _________ and birds after __________
- ♦ _____ ever present (required for natural selection)

Contrast between Evolution and Creation

t 1. Matter ___________
t 2. Sun and stars ______ earth
t 3. Land ______ oceans
t 4. Sun earth's first light
t 5. __________ organisms first, then plants (fish before trees)
t 6. __________ before ______
t 7. ________ and _____ before man

A Biblical view of death

- ♦ "but you must not eat from the tree of the knowledge of good and evil, for when you eat of it you will surely die."
  - ***Genesis 2:17 (NIV)***
- ♦ "For the wages of sin is death, but the gift of God is eternal life in Christ Jesus our Lord."
  - ***Romans 6:23 (NIV)***
- ♦ "For as in Adam all die, so in Christ all will be made alive….The last enemy to be destroyed is death."
  - ***1 Corinthians 15:22, 26 (NIV)***

- “Consequently, just as the result of one trespass was condemnation for all men, so also the result of one act of righteousness was justification that brings life for all men. For just as through the disobedience of the _______ the many were made sinners, so also through the obedience of the _______ the many will be made righteous.”

  -Romans 5:18-19 (NIV)

The Curse on Creation

- Sin of Adam and Eve
- All creation is ____________ (Romans 8:20-22)
- 2nd curse: Destruction by the flood

God judged the world

- “The LORD saw how great man’s wickedness on the earth had become and that every inclination of the thoughts of his heart was _____________. The Lord was grieved that he had made man and his heart was filled with pain. So the LORD said, ‘I will wipe ________, whom I have created, from the face of the earth—men and animals, and creatures that move along the ground, and birds of the air—for I am grieved that I have made them.’” Genesis 6:5-7 (NIV)

Noah’s Flood

- Is there any evidence of a world-wide, global flood?
- ______ graveyards, sedimentary rock, fossil fish in mountains

Genesis gives a historical account

- In the __________ year of Noah's life, on the _________ day of the ________ month—on that day all the springs of the great deep burst forth….And the rain fell on the earth forty days and forty nights." Genesis 7:11-12 NIV
- "And on the _________ day of the __________ month the ark came to rest….The waters continued to recede until the ______ month, and on the _______ day of the _________ month the tops of the mountains became visible." Genesis 8:4-5
- "By the _____ day of the first month of _____________________ year, the water had dried up….By the ___________ day of the _______ month the earth was completely dry." Genesis 8:13-14 (NIV)

Typical Criticisms of Noah's flood

t Where did the water _________?
t Where did the water __?
t Was the ark big enough?

Typical picture of Noah's Ark

Actual Size and Shape of Noah's Ark

Size Comparison for Ark

Can you find the elephant and giraffe?

Was Noah's flood global or ________?

- "For forty days the flood kept coming on the earth, and as the waters increased they lifted the ark high above the earth. The *waters rose greatly on the earth* and the ark floated on the surface of the water. *They rose greatly on the earth* and ___ *the high mountains under the ________ heavens were covered.* The *waters rose and covered the mountains* to a depth of more than _______ feet. Genesis 7:17-20 NIV
- Doesn't leave much that isn't under water

Did the flood really kill *all* of the animals?

- _______ *living thing that moved on the earth perished*—birds, livestock, wild animals, all the creatures that swarm over the earth, and all mankind. ________ *on dry land that had the breath of life in its nostrils died.* ______ *living thing on the face of the earth was wiped out*; men and animals and the creatures that move along the ground and the birds of the air were wiped from the earth. ______ *Noah was left, and those with him in the ark.* --Genesis 7:21-23 NIV
- EVERYTHING! died except what was in the ark (and the fish)

Isn't it ironic?

- The fossil record is a record of _____, when God judged the world with a flood
- Men have ________ this reminder God's judgment against sin into so called evidence of ________—that there is no Creator
- God will judge the world again—with _____: "The heavens will ________ with a roar; the elements will be destroyed by fire, and the earth and everything in it will be laid bare." 2 Peter 3:11 NIV

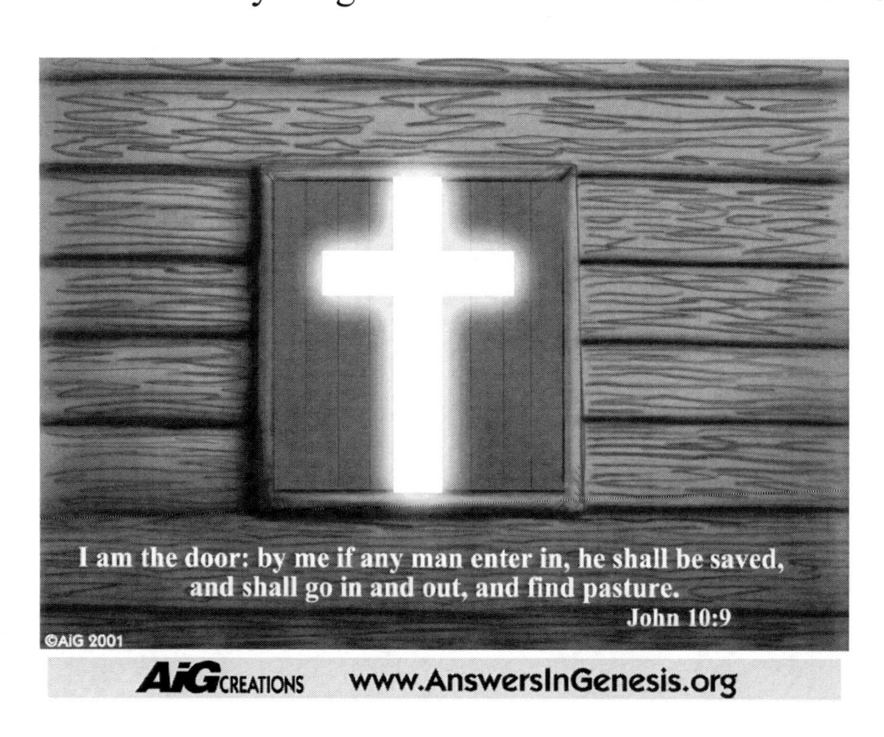

Tyrannosaurus Rex

What did T. Rex eat?

Was it a ferocious predatory hunter?

- "Then God said, 'I give you every seed-bearing _____ on the face of the whole earth and every tree that has fruit with seed in it. They will be yours for food. And to all the beasts of the earth and all the birds of the air and all the creatures that move on the ground—everything that has the breath of life in it—I give every green _______ for food.' And it was so."
  -Genesis 1:29-30 (NIV)

- After the Fall, T. Rex may have been a scavenger but not a hunter

Small arms

"Although all _______ have teeth designed for eating ______, their diet consists mainly of ________."

-Taronga Zoo, Australia bear exhibit,
July 1999

"The oddest inhabitants of the Galapagos coastlines are the marine iguanas, which exist nowhere else in the world. Like miniature dinosaurs, the inky armored lizards swarm over the rocky shores of the islands. When upset they squirt vapor from their nostrils like storybook dragons. Despite their ferocious appearance the sea iguanas are strict and docile vegetarians, completely harmless and gregarious to an extreme. Though armed with strong claws and sharp teeth, they rarely use them on each other and never attack other animals."

Life Nature Library, *Evolution*
Ruth Moore, Time Inc., Pg 21.

Were _______ really dinosaurs?

- Cultures around the world have legends about _______
- These may represent mankind's interaction with dinosaurs in the past
- Remember, the term 'dinosaur' is relatively recent

t "Look at the _________, which I made along with you and which feeds on ______ like an ox. What strength he has in his loins, what power in the muscles of his belly! His tail sways like a _______; the sinews of his thighs are close-knit. His bones are tubes of bronze, his limbs like rods of iron. He ranks first among the works of God, yet his Maker can approach him with his sword.

– Job 40:15-19 (NIV)

Lesson 7

Creation Compromises

Biblical Creation

- Genesis 1-11 represent an accurate, __________ account
- God created the heavens and the earth in the time and manner specified in Genesis: by his ______ in _________.

Compromising Theories of Creation

- Why compromise with evolution?
  - Seem___________
  - don't know scientific evidence against evolution
  - don't understand difference between microevolution/macroevolution
  - don't understand the bible
- May be sincere, intelligent, __________ but uninformed

- Many Christians including pastors and leaders accept ________ of years because of what "________ says"
- Plain reading of Scripture points to a _________ and six day creation

Compromising Theories of Creation

- Theistic Evolution
- Framework Hypothesis
- Progressive Creation
- Gap Theory
- Day-Age Theory
- Old Earth Creation

Theistic Evolution

t Creation by __________ evolutionary process _______ by God: "God used ___________ to create"

Theistic Evolution Contradictions

t _________ the role of God
t ___________ with God's _________
t ________ of creation does not line up
t Biggest problem:
  – _____ before _____

Progressive Creation

t Creation by ______________ evolutionary process by God:
t God _________ only at certain points to "help" evolution along
t But…
t Same but more pronounced problems as Theistic Evolution

Christians Can Have Worldview Conflicts

- "The current theory which I accept points to a big bang theory as the beginning of creation, when about 15 billion years ago…." page 135
- "Whether or not there was a big bang, I don't know. But I am astounded at the wisdom of God to project over 15 billion years the exact mass…" page 157
- "They (Adam and Eve) were still alive, according to my calculation, when Noah came on the scene." page 70 (This is not true if you do the math)

♦ Quotes from *Bring It On* by Pat Robertson

"The 'days' of Creation were ordinary days in length. We must understand that these days were actual days (veros dies), contrary to the opinion of the holy fathers. Whenever we observe that the opinions of the fathers disagree with Scripture, we reverently bear with them and acknowledge them to be our elders. Nevertheless, we do not depart from authority of Scripture for their sake." --Martin Luther

____

(or Ruin-Reconstruction Theory)

Primeval creation existing for billions of years but was _____ ____ in a Pre-Adamic cataclysm

- Some put the fall of Satan here
- Thomas Chalmers, theologian who promoted this view, also G.H Pember
- Advocated in Scofield Study Bible
- Attempt to avoid ________ ________ with _________

Where is the Gap?

- Gap theory proponents place the millions of years between Genesis ______ and ______
- Plain reading does not indicate a gap

Problems with the Gap Theory

- Improper ___________
- Inconsistent with God creating everything in _______ Exodus 20:11
- Puts _____, _______ and suffering before the Fall
- The Biblical pattern is for a _______
- Doesn't really solve ________ _________ and Biblical conflicts

Framework Hypothesis

- t _________ of Genesis 1-11 as true history; it is ___________, ___________
- t "What's important is '____' not '____'
- t But…
- t Biblical figures refer to this as a literal historic account including _____, _____ and _____
- t We trust the Bible for salvation...

Jesus Himself Quotes from Genesis!

- ♦ "'Haven't you read,' he replied, 'that at the ________ the Creator "made them male and female" and said, "For this reason a man shall leave his father and mother and be united to his wife and the two will become one flesh"?'"

  – ***Matthew 19:4 (NIV)***
- ♦ "But at the _________ of _________ God 'made them male and female.'"-Mark 10:6 (NIV)

Day-Age Theory

- t Each day of Creation corresponds to the ______________
- t "A _____ is as a _________…"
- t But...
- t "…and ___________ are as a ____"
- t Plants were made on ________

Attempt to Combine Genesis with Evolutionary Time Scales

- t Hugh Ross' Reasons to Believe
- t Conflicts with both the _____ and evolution!

- The word “day” occurs many times in Scripture
- Only Genesis 1 is in _______ regarding the ________ of the day
- In the past, people were __________ that it took as long as __________

- Even in _______ we have different meanings for the __________
- However, the _______ determines the exact meaning intended

Use of “Day”

- Outside of Genesis “Day” with a ______, _______, ______ or ______ always means a normal 24 hour day
- One exception is _________ (may not be an exception)

t Just in case you missed it, the _______ of Genesis 1 uses normal _______ days

“How long did the work of creation take? When Moses writes that God created heaven and earth and whatever is in them in six days, then let this period continue to have been six days, and do not venture to devise any comment according to which six days were one day. But if you cannot understand how this could have been done in six days, then grant the Holy Spirit the honor of being more learned then you are.”
Martin Luther

Exodus 20:11 (NIV)

t “For in ________ the LORD made ________ and the ______, the ____ and all that is in them, but He rested on the seventh day. Therefore the LORD blessed the Sabbath day and made it holy.”

Old Earth Creation

- Many other hard to classify compromises
- ____________ reasons only for old age
- Genesis 5 is ________
  - Provides ages to show creation is _______
- Gaps are not indicated
  - "______, the seventh from Adam…" Jude 14 (NIV)

Where can millions of years go?

Before Genesis 1:1?

__. Does not help resolve anything.

Between Genesis 1:1 and 1:2?

__. Everything is made in 6 days.

During the Creation Week?

__. Days are normal, solar days

Man is the Focus of God's Creation

- Man reveals God's glory in a unique way through ______
- The Son of the Living God became a ____
- __________: the end of man is also the end of the whole universe

Biblical Creation Merits

- Internally consistent
- Simplest, most direct interpretation
- Upholds truth of _____________
- Only "__________" conflicts with science

Common Criticisms of Biblical Creation Perspective

- How long was a day before the sun?
- Why are the stars so far away?
- Is Genesis 2 a second creation story?

- Hasn't science proven the earth is old?
- What about Noah's ark and the flood?
- Does the age really matter?

- "As for God, his way is perfect; the word of the LORD is __________. He is a shield for all who take refuge in Him."
  ______ __:__

Conclusions

- Your view of origins impacts every aspect of your life

- Many Christians compromise with evolution from lack of knowledge

Lesson 8

Darwin & Natural Selection

Where did all the animals come from?

Before __________, most people believe that God created all living things in ________ the form that we see them today. This is the basis of the doctrine of _______."

London Natural History Museum
Darwin Exhibit, 1997.

Early Foundations of Evolution

_________ (394-322 B.C.) proposed SCALA NATURAE (Scale of Nature)

Ranks non-living and living things (even up to angels) on a "______"

Early Foundations of Evolution

Jean Baptiste _______ (1744-1829)

Initial theory of evolution

- Organisms change through time
- Passed on "________ ________________"

Charles Darwin (1809-1882)

Born to wealthy parents in England

Obtained a degree in ________ from _______

Worked as a _________

Went on the Voyage of the Beagle 1831-1836

Collected specimens in the _________

In the __________

Collected different _____ from the islands he visited
At the time, failed to realize that there were all _______
Darwin shared his poorly labeled collection of birds with biologist
____
_______ found that the jumbled mix of _________, ______, and ___________ are all just different types of _______

Darwin Wondered…

Could varieties of birds, isolated on separate islands somehow turn into ___________?
Why are marsupials in _________?
Why do island animals resemble ________ inhabitants?
________ could select for traits…could _______ select for them?

Darwin Lost His Faith (around ____)

"[D]isbelief crept over me at very slow rate, but was at last complete. The rate was so slow that I felt no distress, and have never since doubted even for a single second that my conclusion was correct."

"I can indeed hardly see how anyone ought to wish Christianity to be true; for if so the plain language of the text seems to show that the men who do not believe, and this would include my Father, Brother and almost all of my friends, will be everlastingly punished."

Autobiography

Darwin was Influenced by

t Principles of Geology by Charles ______

t Essay on the Principle of Population by Thomas ________

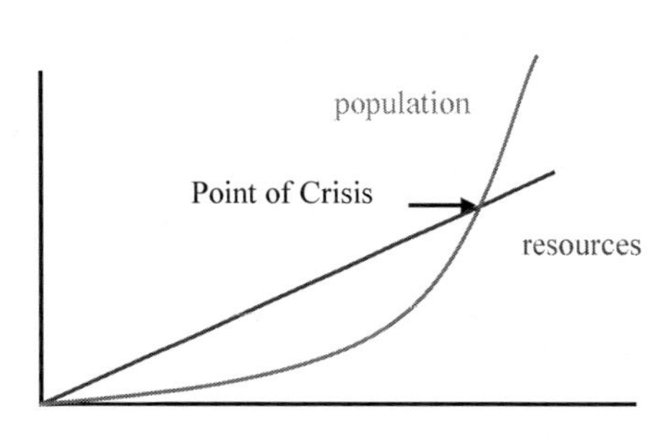

_____ _________

Malthus' Basic Theory

Evolution: Three Components

__________ among individuals

Competition for _________

____

____

______

Heredity

Darwin Wrote *Origin of Species*

Suggested all living things share a ______ __________

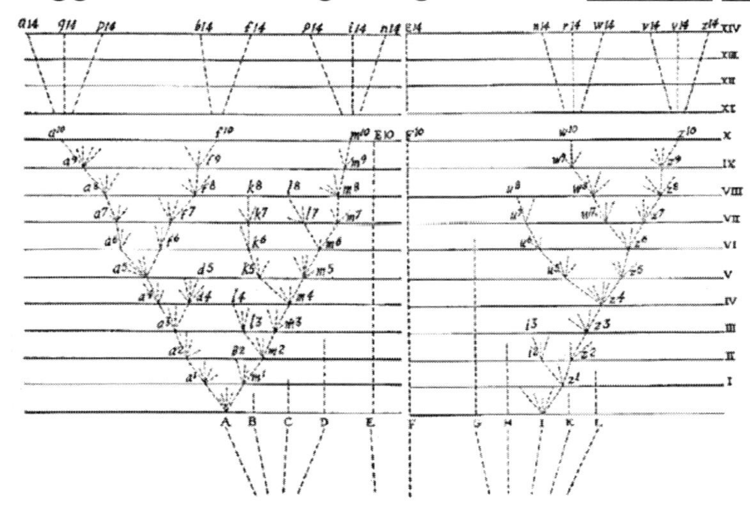

Image from Darwin's book *Origin of Species*

Variation

- t Two organisms can have many different __________ of offspring
- t _________________ leads to _________________

Natural Selection

Environmental conditions favor __________ traits

Individuals with __________ traits have ____ offspring

Individuals ______________ to environment have _____ offspring

Adaptation

___________ organisms tend to lose genetic variability

Specialization reduces ______ adaptability

_______ genetic __________ is a _____ of information

- ♦ The genetic combination that an animal possesses ultimately determines the ______ that it displays or the _______ that it will be

Amazing mixed breed dogs!

Doodleman Pinscher
Labradoodle

Natural Selection is a ______________ Process

Edward ______, Christian and creationist

"intended by __________ to keep up the typical qualities of a ________."

Effect of Mutations

Mutations can generate __________ but the cost is a loss of ___________

Dogs with mutations are still dogs.

Darwin once said, 'The sight of a feather in a peacock's tail, whenever I gaze at it, makes me feel ____!'[26]

Darwin vs. Irreducible Complexity

- "To suppose that the eye, with all its inimitable contrivances for adjusting the focus to different distances, for admitting different amounts of light, and for the correction of spherical and chromatic aberration, could have been formed by natural selection, seems, I freely confess, ______ in the highest possible degree…"

  – ***Charles Darwin,* On the Origin of Species**

Darwin vs. Irreducible Complexity

- "If it could be demonstrated that any complex organ existed which could not possibly have been formed by numerous, successive, slight modifications, my theory would absolutely __________."

  – ***Charles Darwin,* Origin of Species**

"numerous, successive, slight ____________"

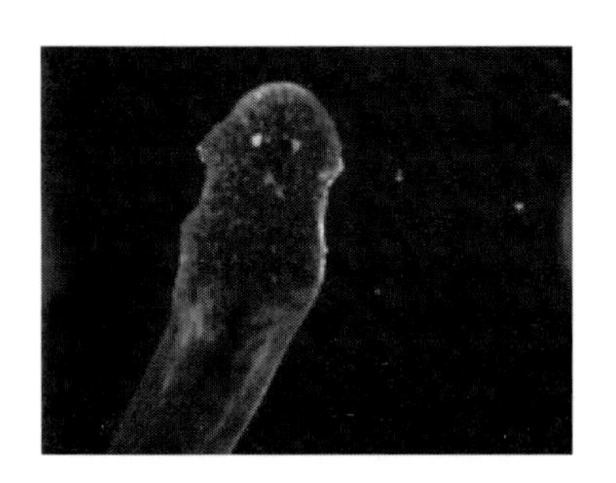

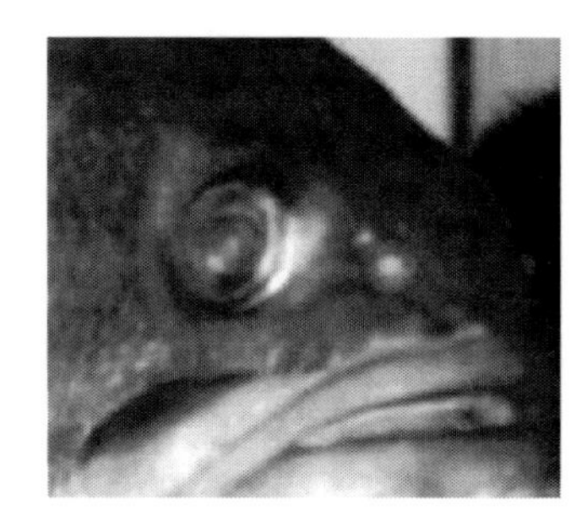

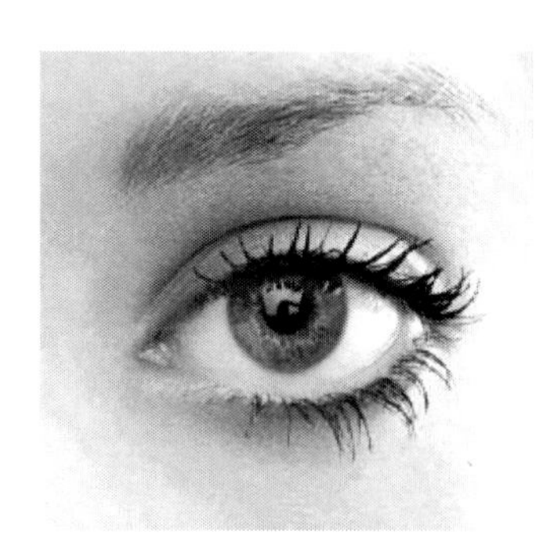

Darwin vs. Gaps in the Fossil Record

As Darwin himself pointed out in his book, The Origin of Species:

". . .[T]he number of intermediate varieties, which have formerly existed on the earth, [must] be truly ________. Why then is not every geological formation and every stratum full of such intermediate links? _______ assuredly does not reveal any such finely graded organic chain; and this, perhaps, is the most obvious and gravest ________ which can be urged against my theory."

Micro-evolution vs. Macro-evolution

- Scientific evidence supports change within a kind, __________
- No scientific evidence to demonstrate change into a__________ _____

Lessons 9 & 10

Genetic Limits to Evolution

Evidence for Evolution

1. Microevolution--variation within kinds
2. ___________--Morphology, embryology, biochemistry (DNA)
3. "Mistakes" or ________ of evolution
4. The ____________ (most important)

Vestiges of evolution?

♦ Examples of "vestigial" organs
  - __________
  - __________ in snakes

♦ vestigial organs ___ have functions

Similarities (Homologies) Are Only Circular Evidence for Evolution

♦ "Since all members of a taxon must consist of the descendants of the nearest ___________, this common descent can be inferred only by the study of their _____________ characters. But how do we determine whether or not the characters of two species are ____________? We say that they are if they conform to the definition of homologous: *A feature in two or more taxa is _____________when it is derived from the same (or a corresponding) feature of their nearest _______________.*"

– Ernst Mayr What Evolution Is 2001 p. 16 (emphasis in original)

bara min

t _______ for created kind

t "______ of species" vs. ______ of kind

An incorrect creation view where organisms remain ________ the same

Creation

Time

Morphology

Microevolution: Variation Within Kinds

Mongolian Horse

Shetland

Donkey

Zebra

Zonkey or Zedonk

Microevolution Vs Macroevolution

Microevolution

Changes in the ______________ of a _________ over time

Modification of ____________ genetic _________ in population

Macroevolution

Addition of ___ genetic information in a population

Large-scale changes linking one "kind" to another.

Examples:

Dinosaur/Bird   Reptile/Mammal   Terrestrial Mammal/Whale

Macroevolution—Increase in genetic information, common ancestors

Microevolution—variation within the group, independent origins

Factors contributing to Microevolution

- Genetic _____
- _________
- Gene _____
- ________________

Homologous Chromosomes

- Contain _____ that code for the same ______
- May contain the same or different alleles
- _______ are different forms of the same trait
- Gene _____: location of gene on _____________

A homologous chromosome pair
showing homozygous & heterozygous alleles

Image by Daniel Howell

Genetic Drift

- ______ loss of _______ or traits in a population through time.
- ___ impact on _____ populations.
- ____ populations are more ___________.

Mutations

- ___________ changes in ______
  - Harmful
  - ________
  - Beneficial (heterozygote, resistance)

    So rare there are virtually no real examples
- Most occur when DNA is __________

- Occur without reference to ______ ____________
- Occur in organisms _______ adapted to their environment

Gene flow

- Flow of _______ between ___________
- Occurs through _________/____________
- Alipatric speciation
  - New species are supposed to arise through ____________ isolation (new species but not ____ ______)

Natural selection

- Genetic variation and ______________
- _____________ conditions "select" traits in a population most suitable for ________ (natural selection)

Darwin's Observations

- _________ exists in all natural ___________
- _________ beyond capacity of _________
- Resources are _______
- Most advantageous variations ________
- Natural selection is a _______ process

Classic Examples of Natural Selection

- Galapagos finches
- Variation with selection not _________

Classic Examples of Natural Selection

- _____________
- Change in % dark colored moths after the ________________
- But…Moths don't rest on tree trunks!

Natural Selection?

Where did the giraffe's long neck come from?

Other examples

- t Domesticated animals, crops
  - --dogs
  - – 4 legged chicken
  - – tomatoes, corn
- t Insecticide resistance

Antibiotic resistance in bacteria

Bacterial antibiotic resistance results from ____ and mis-regulation

Evolutionary _________?

- t Some "species" are not really separate species but _______
  - – Red wolves
  - – Northern spotted owl
  - – Macaque monkeys in Indonesia
  - – Ducks and warblers
  - – Cutthroat and rainbow trout
- t Appears to be the ________ of evolution

Microevolution

- t Variation within species is evidence of the variety ________ within each ______

Other factors

- – bottlenecks
- – founder effect
- – _____ populations_______ change
- – _____ populations risk extinction

Population Bottleneck

Reduced Genetic Diversity

Macroevolution

- Addition of ___ features within a population when the ancestral stock neither had these features nor the ____________________ which codes for it.

Evolutionary Explanation for Source of New Genes

However, most genes are "orphans" with no known origin

Types of Macroevolution

- Neo-Darwinism
- ________________

Neo-Darwinism

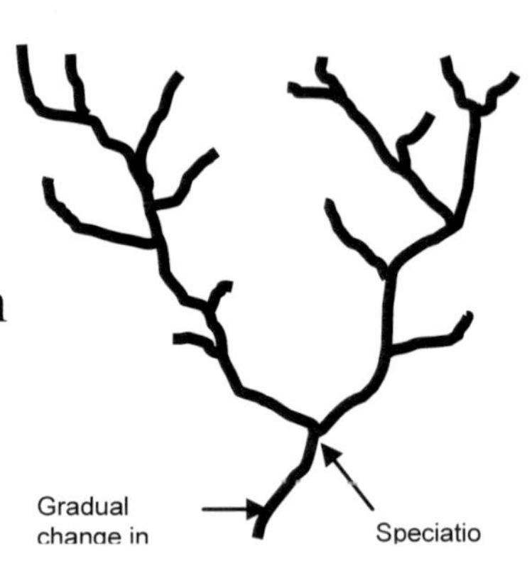

- Evolutionary change is a ____________ process evenly distributed over _____. Origin of novel adaptive structures is through ________ and ______________.

Punctuationalism

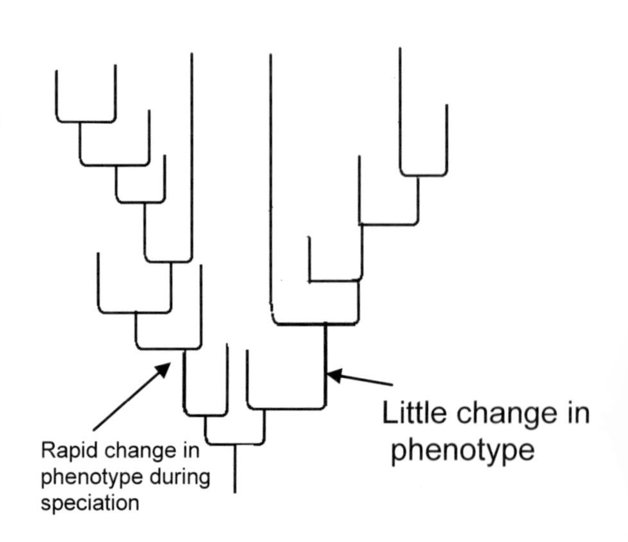

- Evolutionary change in ______ spurts with long periods of ______, short periods of _____ change. Rapid appearance, fully formed, with __ apparent ancestors.
- Origin of new traits is ________ through _________ and natural selection
- Used to explain the ___________

Difficulties with macroevolution

- Duck billed platypus
- Dolphin/whale
- ________
- Feathers

Where does the ***NEW*** genetic information come from?

Feather structure

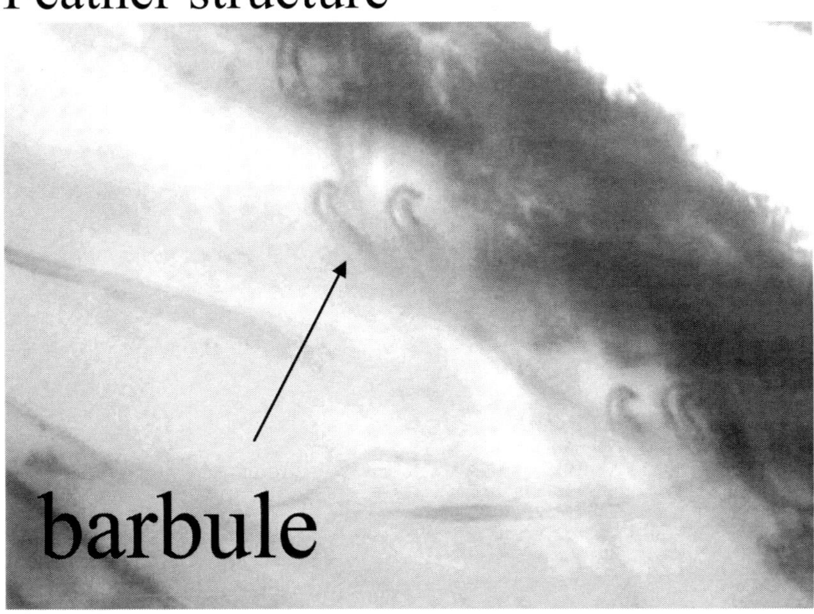

Further...

- Mutations are mostly _______ or neutral, very rarely beneficial
- Mutations represent modifications of ________ genes!
- How can mutations lead to new traits?

How are people deceived?

- Given plenty of strong evidence for ______________ and ________________
- Evolutionists _____ the __________ between micro- and macro-evolution

Notes

Lesson 11

How old is the Earth really?

Why does age matter?

t Some reject Christ because they consider the Bible a book of __________ not _________
t If we cannot trust the Bible when it speaks of "_______things" why believe what it says about ________ things
t This is not an esoteric scientific or theological debate but a relevant one to all

How can we *know* how old the earth is?

♦ __________
  – Does the scientific evidence demand an ___ earth?
♦ __________
  – Does the Bible require a _____ earth interpretation?

Major types of scientific evidence suggesting an old age for the universe

1. Distant ________
2. Big bang theory
3. ______ record
4. __________ dating

Distant starlight

♦ Stars are _________ of light years away and thus it would take ________ of years for the light to get here
♦ ___________ explosions indicate that light cannot have been made "___ __________"
♦ Speed of light may not have been ___________
♦ _______ affects the speed of light and also _____, therefore ______ ___________ may be a factor

"Big Bang" Theory

- All ______ and ______ in the universe was once in a single dimensionless ______ of infinite _______
- Suddenly, this ___________ into the present universe
- This event supposedly took place _____________ years ago

Evidence for the "Big Bang"

- ___ _____ of spectra from galaxies are interpreted as _________

Alternative explanation for red shift

- According to the "big bang" theory, the universe has _________ and __________
- Model is galaxies on the surface of a _________

Alternative explanation for red shift

- However, _________ red shifts raise the possibility that galaxies are positioned in "______" with the Milky Way close to the _______ of the _________

Another Big Problem for Big Bang

Far out in space, scientists believe they are "looking back in time"
However, there are "_______ galaxies" as far as we can tell

Fossil record and extinctions

- Large number of extinct animals
- _______ of fossils in rocks suggests __________
- ____________ are extinct and supposedly died out ~____________ ago

Challenges for the fossil record

________ _____________

Organisms appear with no ancestors

"_______ fossils"

General _______ of transitional fossils

_____ tissue from T. Rex dinosaur

Radioisotopes with long half lives suggest an old age for the earth

Scientists measure the ______ of parent/daughter isotopes
Based on the isotope _____ and the ________, the time is inferred
_______ are thus assigned to rocks and fossils

Radioactive Isotopes

t Amount of "parent" decreases exponentially as
t the amount of "daughter increases exponentially
t a half-life is the length of time for 1/2 of a radioactive element to decay

Types of Radiometric Dating

t Uranium--->Lead
  – U^{238}--->Pb^{206} 4.5 billion years
  – U^{235}--->Pb^{207} 0.7 billion years
  – Th^{232}--->Pb^{208} 14.1 billion years
♦ Potassium-->Argon 1.3 billion years
♦ $Rubidium^{87}$--$Strontium^{87}$ 60-120 billion years
♦ $Carbon^{14}$ -->$Nitrogen^{14}$ 5,730 years

Assumptions required for Radiometric Dating

t 1. The system is closed
  – No addition and no loss
t 2. No daughter present at formation
  – presence of daughter would increase age estimate
t 3. Decay rate has been the same

Problems with radiometric dating

Carbon 14 found in 300 million year old _____ but carbon's half life (~______ years) means there should be ______
Carbon 14 also found in __________
These and other evidences suggest that the decay rate has
____

Does the scientific evidence *demand* an old age for the universe?

1. Distant starlight ____
2. Big bang theory ____
3. Fossil record ____
4. Radiometric dating ____

Does the scientific evidence *demand* an old age for the universe?

- While much of the scientific evidence may be _____________ to support an old age for the earth and universe, it is ____________ and there are growing reasons to doubt
- Most of the evidence ___________ with an old age is _____ and some comes from ____________ sources

- "…I am a young-age creationist because that is my understanding of the Scripture. As I shared with my professors years ago when I was in college, if all the **evidence** in the universe turns against creationism, I would **still be** a creationist because that is what the Word of God seems to indicate. Here I must stand." Kurt Wise

  – ***Six Days: Why 50 Scientists Choose to Believe in Creation***

Does the Biblical evidence *demand* a *young* age for the universe?

- A straightforward reading of the Bible strongly implies a young age for the earth
- A young earth model is _________ _________ and makes sense of the Biblical data
- Most old earth views ignore or force __________ ___________ of Scripture
- The motivation for nearly all old earth theories is to accommodate ___________ ______

Two Separate Issues Regarding Age

- Length of days in Genesis 1
  - Six ~24 hr days or long periods of time
- Age of the earth and universe
  - Millions of years could be before the six days which are recent
  - Millions of years could be after the six days

Biblical issues about age of the earth

1. Genesis 1 is ________
2. _____ ______ in Noah's day
3. Long lives of _________
4. One man—Adam
5. Words of _______

Genesis 1:1-2:3 is historical narrative not poetry

Steven Boyd performed a statistical analysis of verb tenses
Genesis 1:1-2:3 has verb tenses consistent with ________ ________ not ________

Millions of years cannot be placed during six days of Genesis 1

- Genesis 1 is historical narrative
- There was ________ and __________
- Exodus ___:____
- Establishes the pattern of the _____ ________
- Virtually all Hebrew scholars recognize that the intent was to convey six _______ _____ of ______ length

Where is the evidence for the global catastrophe?

Much of the ______ ________ was formed during the flood and shortly thereafter with flood associated events

Thus, the geologic column ***EITHER*** provides evidence of a global flood ***OR*** it is evidence for the millions of years. It cannot be both.

Did the patriarchs really live so long?

- Current cell & molecular biology research cannot rule out the possibility
- _________, a disease resulting in rapid aging, could be a model
- Perhaps we all suffer from rapid aging compared to __________

How old is the earth…really?

Current scientific evidence leaves open the possibility of a young age for the earth and universe

Plain meaning of Scripture implies a young age for the earth and universe

Therefore, it is more than reasonable for a Christian to take a young earth view.

Scripture should inform our interpretation of scientific evidence not the other way around.

Lecture 12

Thousands not Billions

Notes

Notes

Lesson 13
Origin of Matter & Energy

Evolution Model

- t Must account for:
  - – _________, diversity
  - – origin of life (________)
  - – origin of the ___________
  - – origin of matter/energy/universe

Creation Model

- t Already accounts for all of these
  - – God made all life to reproduce ______________
  - – God said "Let there be…" and there was
  - – "The heavens declare the glory of God"
  - – See Job Chapters 38-41

Matter and Energy

- t Matter has ____, occupies _____
- t An ______ is the smallest particle of an element with the ________ of that ___________
- t _________ consist of two or more _____ bonded together.
- t ______ is the capacity to do _____, make things happen

Origin of Matter/Energy

- t Creation Model
  - – Matter, energy, laws called to exist by a Creator. Complete, complex, conserved but _______ __ _______ since the fall

- ♦ Evolution Model
  - – Matter, energy, laws came from ________ to form subatomic particles (plasma) and the complex forms of matter (atoms, molecules, stars)

What do we "know" about the universe?

- _______________ reflect what we know about the universe
- First law of _______________
- Second law of _______________

First Law of _______________

- Conservation of _______/_______
- _______/_______ can neither be _______ nor _________, but can be _______ into one form or another

Second Law of ___________

- "Everything goes from bad to worse"
- Every ______ left to itself goes from _______ to _______. ______ is transformed into less useful forms (_______).
- In a spontaneous process, _______ _________.

Evolutionary Theories on Origin of Matter

- __________ Theory
- _________ Theory

Both theories are inconsistent with the laws of thermodynamics

- _______________ theory generally _______ now
  - Inconsistent with first law of ___________
- _________ theory points to a "_________"
  - Difficulty with initial ordered state
  - The Universe is "______"

Evidence for __________

- t Most popular natural explanation
- t Major evidence for thc ________ theory is evidence that the universe is _________
- t BUT…
- t An ________ universe is still consistent with biblical creation since the Bible says that God "_________" the ________.

Origin of the Solar System

- ♦ 1600's Rene ________ suggests initial material in universe rotated like a whirlpool
- ♦ Mid 1700's ______ thought a rotating cloud of gas would contract in the middle and flatten
- ♦ Later 1700's ______ proposed cooling, shrinking, rotating cloud would leave rings that would eventually become planets
- ♦ ______ suggested the sun was hit by a comet and pieces flew off to form the planets

Nebula Model (Evolution)

- ♦ Heavy elements produced in stars and released through _______________
- ♦ Solar systems form from _______ of gaseous material in a __________
- ♦ _________ and angular _________ produce spinning disk with central _______
- ♦ _____ material condenscs to form ________

Evolution Model

- t ____ should be spinning _________
- t Material of planets should be _________
- t Orbits in ________ plane
- t No retrograde __________

What do we know now?

- Most (___%) angular momentum is in the planets
- Mercury, Pluto, comets, asteroids are not in orbital ______
- ______, _______, and 3 of 12 _______ moons have ________ rotation

- "Of course, there are many more details to this story that we have not considered. Some of the planets have gaseous atmospheres and some do not. Some have satellites, like our moon, and some do not. Some planets rotate in the same direction that the sun does and some rotate in the opposite direction. Any theory of the beginning of our solar system must explain all of these observations. Many questions remain to be answered before a complete explanation can be made."
  - GENESIS "Exploring Origins" science module www.nasa.gov

Creation Model

- Each celestial body was individually created and therefore they do not need to be similar in ________, orbital _____, or ________

Origin of the Moon

It formed as ___________________ and separated
  - _________ __________ is a problem
  - Moon lacks iron/nickel core like ________
- It formed ________________ and was __________
  - Similar __________ isotope ratios is a problem
- Some suggest Mars-sized body _________ with the early earth to create the ______
  - Some models suggest such an impact would preclude liquid ________ on _________
- Earth is the only planet in the solar system with _____ and ____ ___________

Conclusion:

- No good naturalistic explanation for the origin of the solar system

No experiments (scientific method) can be done to prove it anyway

Lesson 14
Fossil Record 1

The fossil record is one of the most significant areas in the creation/evolution controversy

Fossils: _____ that ________ lived in past

- Types of fossils
  - Preserved _____ ______
  - Casts/Molds (external shape)
  - _______ (footprints, tunnels)
  - ______ (insects trapped in sap)
  - _________

Yes, there is fossil _____!

Formation of Fossils Requires:

- Rapid _______ of organisms
- Rapid __________ of sediments

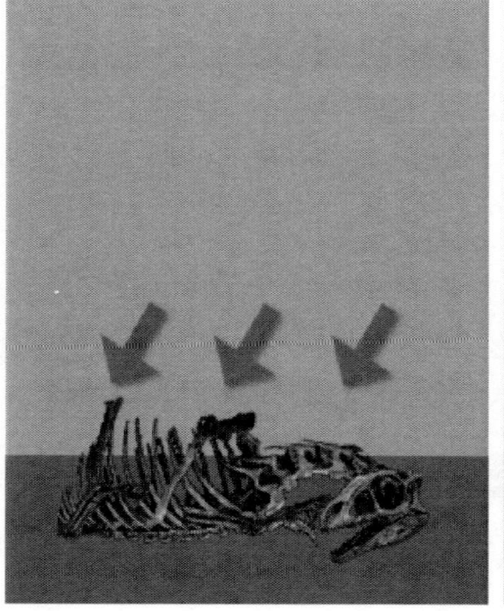
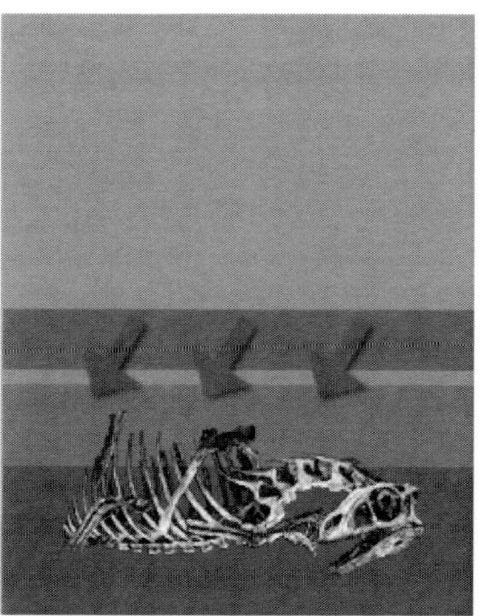

- Rapid ______ is required because organisms must escape ________ and ___________

Formation of Sedimentary rock

- 1. _______ and __________ in layers

- 2. ________: sediment changed into rock
  - Cementing agents (_______)
  - Compression/__________

Fossil Record Predictions (Evolution)

1. Gradual sequence from ______ to _______ organisms
2. _______ ______ should be ________
3. Living species significantly _________ than fossil ancestors
4. Modern species __________ be found in rocks older than rocks containing supposed ___________

Fossil Record Predictions (Creation)

1. _______ appearance of _________ organisms with

____

2. Organisms remain ___________ the same to _______ or become _______
3. Fossilized modern species _____ be found in rock containing extinct species
4. Different __________ of fossil species in different ______ indicate ecological ______ not evolutionary history.

History of Evolutionary Thought Regarding geology

until mid to late ______ common belief was that sedimentary rocks/fossils were all deposited in _______ ________

_____________--rapid, violent, major events could radically shape the planet very quickly

_________________– "the present is the key to the past"

_________ views were challenged by the __________ Geologists

____

- ______ and others believed a ______ of _________ was supposedly responsible for geologic formations
- He thought God supernaturally ________ the earth after each _________
- Did not take a ___________ approach to Scripture

Cuvier

____

- Only processes ________ _____ can be used to describe ____ events
- Proposed by James _______, popularized by Charles _____
- First theory to suggest that the earth was very _____.
- Viewed earth and universe as continuously changing with no __________ and no ______

_______________ Geologists

- Group who objected to ____ ________ compromises
  - Some Christians had compromised (Gap or day age)
- ___________ objections
  - Old earth geologists ignored Genesis 6-9 (Noah's flood)
  - An old earth postulates long ages of violence and death
- __________ objections
  - Deposition of sediments was _______
  - __________ fossils
  - Some fossil human remains were found with extinct animals of a supposed earlier geologic age
  - Geology was really in its ______

Some Strata are Tightly Folded

Photo from the Grand Canyon courtesy of Tom Vail

___________ Fossils

- Some fossils exist through more than one strata
- Example: fossilized or petrified _____ which cross strata

Other players

- William _____: Principle of ________ _________
  - Fossils increase in ________ with time and fossils could be used to ________ (date) strata around the globe
- Charles _______: Theory of natural ________ (origin of species through time)

____________ _____________

- Different types of organisms are found in different types of ______
- Increasing _________ occurs more _________ in geologic time (not really)

Fossil Predictions

Creation model	Evolution model
_________________ and fully formed	____________________ from simple to complex
__ transitional forms	________ transitional forms

Gradual or Sudden Appearance?

- "Oldest" rocks containing abundant fossils are from the ________ period
- The ________ _________ is "Evolution's Big Bang"
- ______ appearance of a wide variety of living things with no apparent ancestors

- Time Magazine: "New discoveries show that life as we know it began in an amazing biological frenzy that changed the planet almost overnight."

Cambrian Fossils 543-510 mya

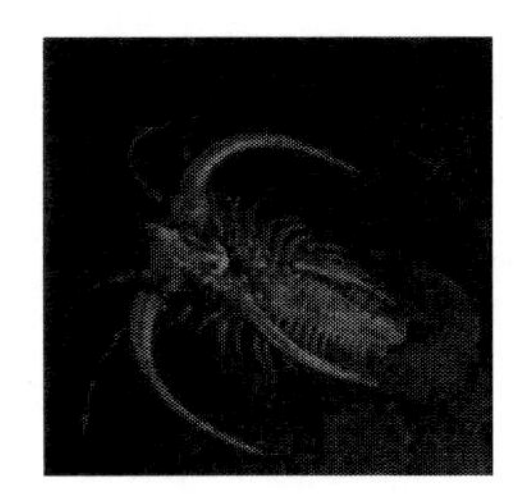

Marella

Anomilocaris

Eldonia

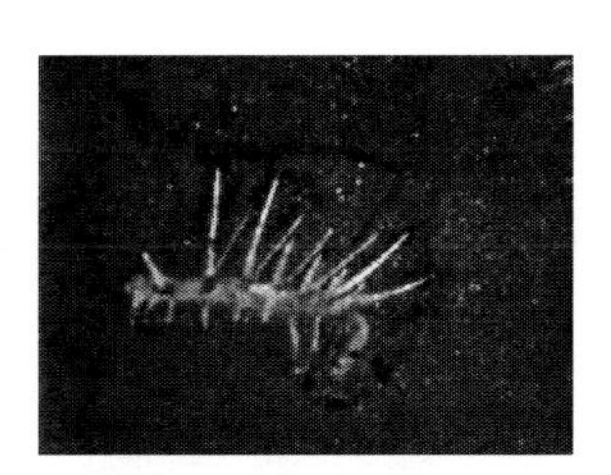

Halucagenia

Lesson 15

Fossil Record 2

What is the evidence for evolution?

Often claim "evidence is ____________"

Only "_________" evidence can be used

Three main lines of evidence

- ___________
  - Similarities
- _____________
- ________ and _________

Macroevolution: Large Changes

- Only "observed" in the ______ ________
- Evolution of mammals from reptiles
- Evolution of Whales
  - beautifully documented
- Human evolution
  - "beautifully documented"
  - beginning with Australopithecine
- What is the evidence?

Reptile to Mammal Evolution

- "No fossil amphibian seems clearly ancestral to the lineage of fully terrestrial vertebrates (reptiles, birds, and mammals)."
  – ***Stephen J. Gould,* Natural History *100:25***
- ________ and alleged evolutionary descendants (synapsids) appear at the same time
- Some specimens are known only by _____ and ___ fragments

- No fossil evidence for origin of modern mammals:

monotremes __________ placentals

Evolution of Whales

t *Evolution of Whales?*

Whale Evolution

- Many of the alleged intermediates appear only as ____________
- Some are based on very ______________ evidence

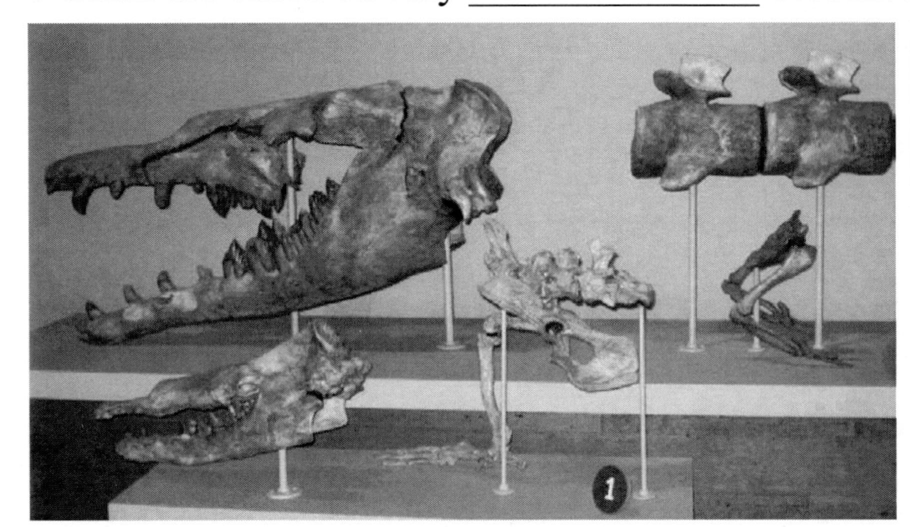

Fossils at Field Museum in Chicago

Front is Rodhocetus,
back is basilosaurus

Evidence for Whale Evolution?

"In time and in its morphology, _________ is perfectly intermediate, a missing link between earlier land mammals and later, full-fledged whales."

Phil Gingerich, "The Whales of Tethys," *Natural History,* April 1994, p. 86.

__________ – 7 years later (2001)

"terrestrial mammals, no more amphibious than a tapir."

Based on fossil evidence discussed by J.G.M. Thewissen, E.M. Williams, L.J. Roe, and S.T. Hussain, "Skeletons of terrestrial cetaceans and the relationship of whales to artiodactyls," *Nature*, 413 (20 Sept. 2001), 277-81.

"Human Evolution is Beautifully Documented"

"It doesn't work very well"

____

"The history of most fossil species includes two features particularly inconsistent with ______________.

- _______: Most species exhibit no ___________ _______ during their tenure on earth. They appear in the _______ _______ looking much the same as when they disappear; morphological change is usually _______ and __________.
- ______ _________: In any local area, a species does not arise __________ by the steady transformation of its ancestors; it appears all at once and '_____ _______.'"

– ***Harvard Professor, Stephen J. Gould, Ph.D.***

"_________ Fossils"

- Some organisms remain ________ through "millions of years of evolution
  - ___________
  - Horse-shoe crab
  - ______
  - Coelocanth
  - Ginko biloba

Evolution of Vertebrates

- _________ present in "lower" layers
- ______ present "upper" layers
- Transitional forms?

Fish—Many different "Kinds"

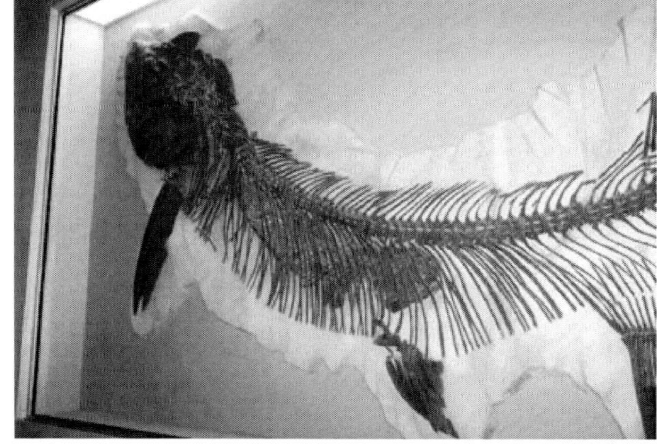

- Placoderms—extinct armored fish
- ________—sharks and rays
- ___________—bony fish (most modern fish)

Chondrichthys and Osteichthys

Tetrapods—Four-limbed vertebrates

- Amphibians are supposed to have evolved from fish.
- What do the fossils show us?

Reptiles and Birds

Archaeopteryx

- Supposed intermediate between dinosaurs and birds
- has feathers
- has teeth
- has claw on wing

- This is a <u>bird</u>.

Other mosaic birds

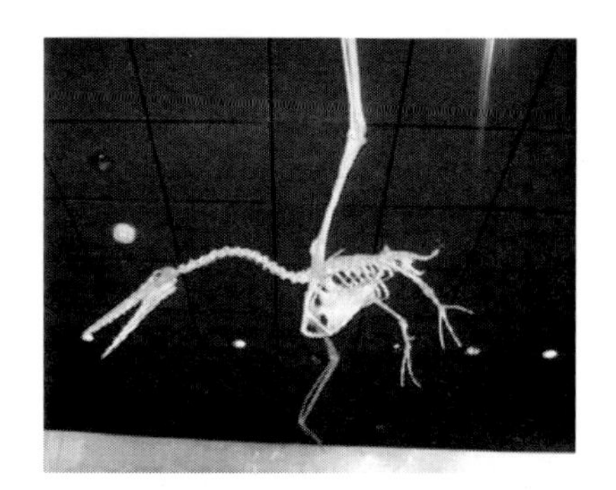

Pseudodontorns

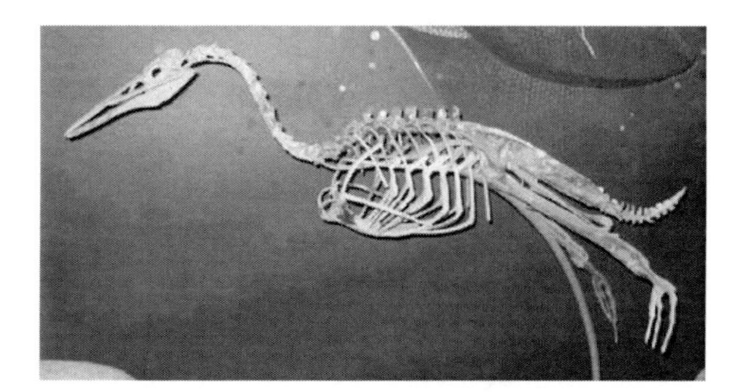

Hesperornis

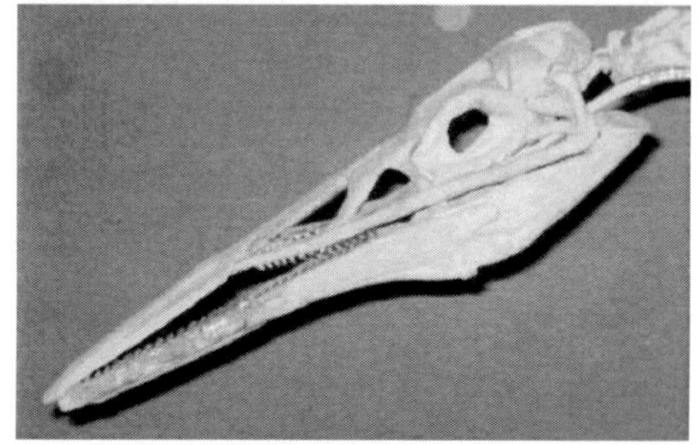

Fossil birds had teeth.
A few modern birds have claws

Fossil Record of Mammals

- Many different types of mammals throughout the rock record.
- More modern groups first appear in rocks of "__________" age
- May represent post-Flood _____________.

Fossil Record of Mammals

Evolutionary History is _________!

- "The extreme ______ of ________ ______ in the fossil record persists as the trade secret of paleontology. The evolutionary trees that adorn our textbooks have data only at the ____ and _____ of their branches; the rest is ________, however reasonable, not the _________ of _______."
  - Dr. Stephen J. Gould, Harvard Professor

Are there transitional forms?

- Depends on what one means by "transitional"
- An animal can be intermediate in **form** between two organisms, without being a **genealogical transition** between them.
  - Ex.: duck-billed platypus (reptiles and mammals)

Are there transitional forms?

- There are genealogical transitions ______ each created kind, but not ________ the created kinds.
  - Ex.: there must have been transitions linking wolves and dogs, or lions and tigers.
  - *But*, there are **no transitions** between the dog and cat kinds.

Are there transitional forms?

- Overall, there are __________purported genealogical transitions in the fossil record.

- Most groups make their first appearance in the fossil record __________, and with no clear ____________________.

Evolution's answer

- ________________--
- Long periods of ______ (no ________), short periods of ______ ________
- Proposed by Stephen J. Gould and Niles Eldridge

Which model fits the data best?

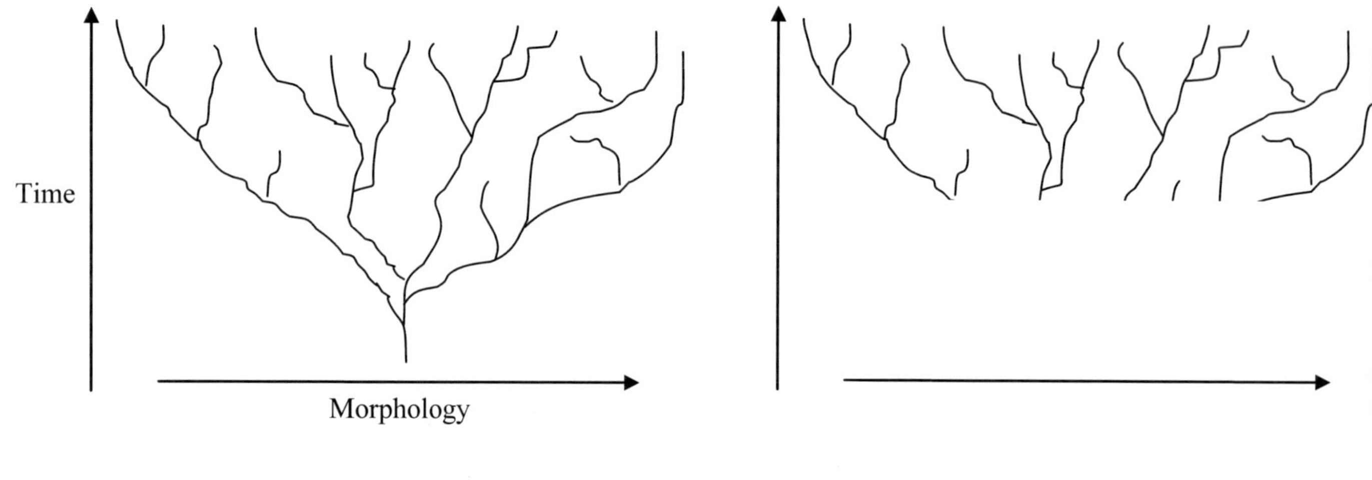

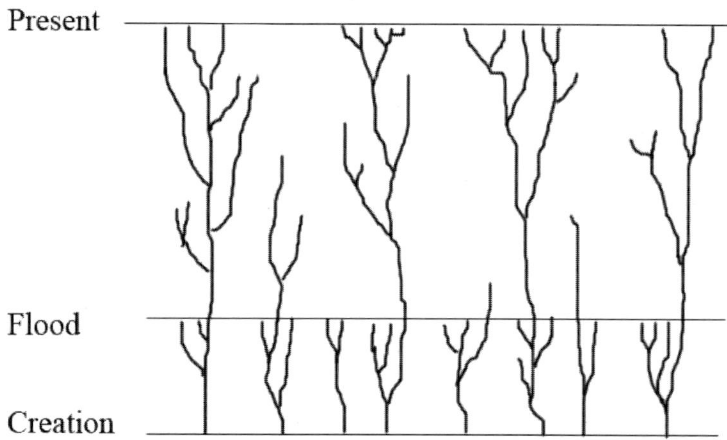

There is a Creator!

- If there is a Creator then there is someone to whom we are accountable.
- God has created and redeemed us through Jesus Christ

Lesson 16

Fossil Record 3

Notes

Notes

Lesson 17
Human Origins 1

Evolution model predicts a ______ ________ of man and apes.

In Scripture there is only one race: The _____ Race.

"Adam named his wife Eve because she would become the mother of all the living." Genesis 3:20

UV vs. skin reflectance

Parts of the world with the most UV exposure are areas where ____________ people have the __________ skin
Parts of the world with the least UV exposure have people with the ____________ skin
UV is required for the production of __________, but destroys folic acid, an important _______________

Genetic Studies

t Based on ____________ analysis or mitochondrial DNA
t All males descended from a common "Adam" about 100,000 years ago. (Likely much more recent)
t Everyone descended from a common "mitochondrial Eve" about 200,000 years ago. (Likely much more recent.)

Do humans and chimps have >98% DNA similarity?

Recent studies have _______ this number
– Insertions/deletions account for more differences
Humans have __ chromosomes, chimps __
Human ____________ are ____ as long
Many other differences not quantifiable as percentage
Conclusion: Humans and chimpanzees ______ share a common ancestor.

AGTCGTACC→AGTCATACC
Substitution

AGTCGTACC↔AGTCTACC
Insertion/deletion

Human and Chimpanzee Chromosomes are different

- Evolutionists suggest progression because of _____ and "transition from ape-like to human-like characteristics:
  - transition to _______________
  - increasing _____ size

Alleged Progression of Human Evolution (out dated)

t Old world Apes	_____ m.y.a.
t Australopithecus	____ m.y.a.
t Homo habilus	_____ m.y.a.
t Homo erectus	_____ m.y.a.
t Homo sapiens	
– Neanderthalensis	35,000-200,000
t Homo sapiens sapiens	34,000 to present

Major Skeletal Differences (For walking)

- Human spine: ____ and _______
- Pelvis:________
- Thighbone is angled ______
- _____ joint
- ______ is ________ and no opposable big toe

Some differences between human and chimpanzee skulls

- Chimps have:
  ______ face
  _______ braincase
  Large, long jaw
  No _____ bones
  Large __________ teeth
  _____ zygomatic arch
  Brow ridge

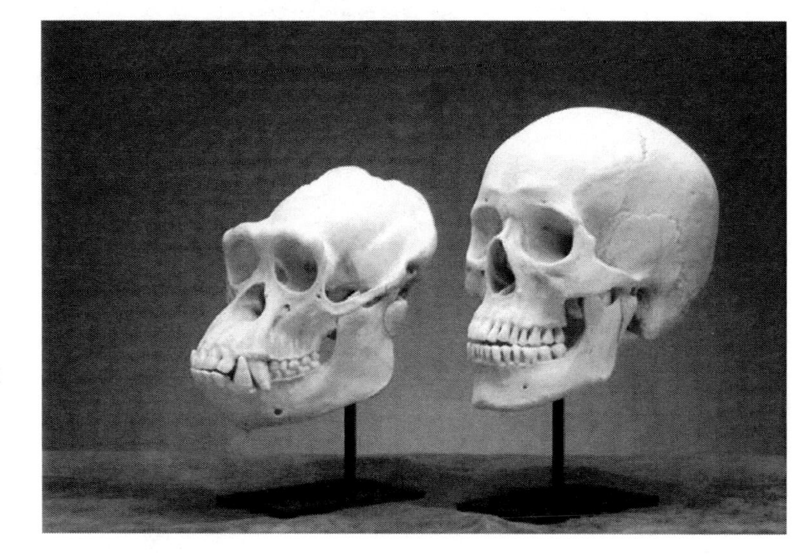

Human jaws are different from the apes

- A) orangutan
- B) gorilla (M)
- C) gorilla (F)
- D) chimpanzee
- E) human

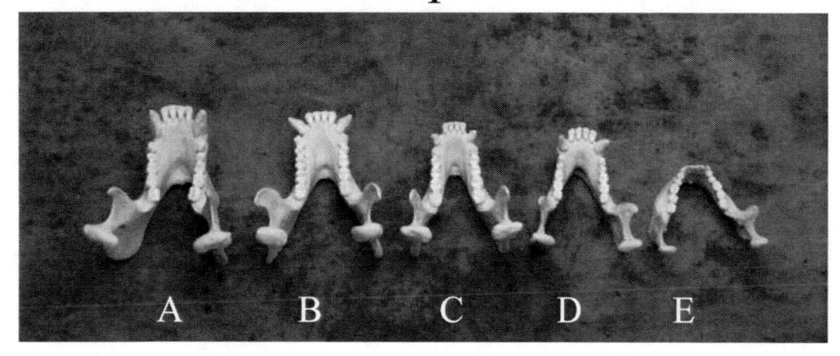

Quadruped vs. Biped Skull

- A) orangutan
- B) gorilla (M)
- C) gorilla (F)
- D) chimpanzee
- E) human

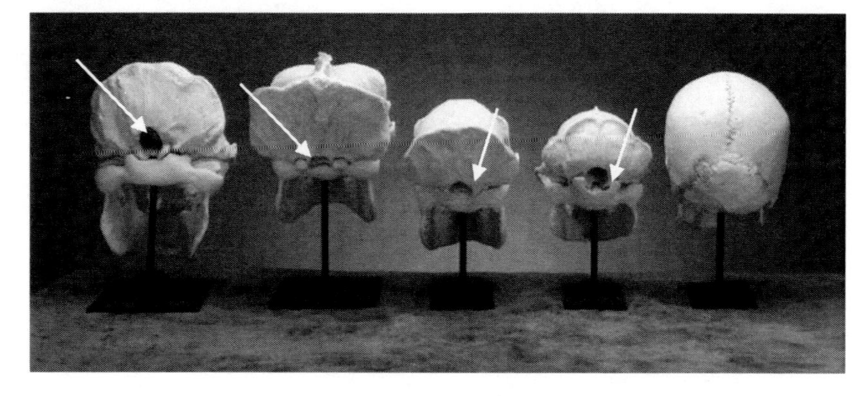

Ape skulls are resting on a white platform base

Location of ________ _________ is toward the rear in qudrapedal apes

Our Oldest Ancestor?

- July 2002, researchers reported the "______" skull from ____
- Dated 6-7 million years based on associated fossils
- Surprising features, ape sized brain, but _____ face and _____ canines "mixture of primitive and advanced features"

Just wait a little while…

A few ____ later, other scientists dispute the claim.
Features they said were "human-like" really are characteristics of

______ ________

Was __________ bipedal?

Brunet has argued that the creature walked on two legs. However, the position of the _______ ______ (arrow) is located far to the rear just like the apes that walk on all fours.

"Proves" is a strong word

"the A. r. kadabba toe bone has a humanlike upward tilt to its joint surface, but the bone is long and curves downward like a _______ does (which somewhat obscures the joint's cant)." --*Scientific American* **13**(2):8

"To me, it looks for all the world like a __________ foot phalanx" comments David Begun of the University of Toronto…."
--*Scientific American* **13**(2):9

Evolutionists often portray organisms in such a way as to make them appear in a progression.

Apes can be ________ or humans __________ to make them appear more similar

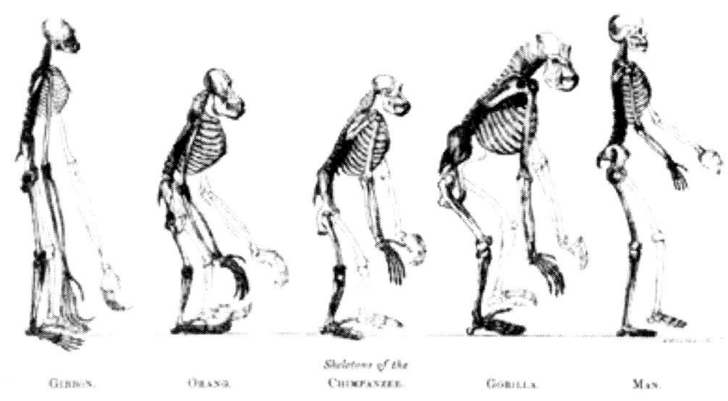

From Thomas Huxley

How similar are humans and chimps?

____

Brain Capacity

- t Gibbon 97 cc
- t Chimpanzee 400 cc
- t _______ 543 cc
- t Australopithecine 650 cc
- t Homo erectus 700-1200 cc
- t ___________ 1250-1650 cc
- t Modern humans 700-2200 cc (1250 ave.)

The _______ Footprints

- ♦ Footprints left in ________ ___
- ♦ At least ___ individuals walking side by side, walking upright with no diverged big toe
- ♦ Presumed to be ______________ because of the 3.6 m.y.a. date

♦ However, there is still debate whether Lucy walked ______, was ________ or _________ walker

"What does it mean to be human"

t Man is created in the image of God

t God so loved the world that he sent his only begotten Son

Lesson 18

Human Origins 2

Human Evolution?

"Despite widespread searching, diagnostic fossils of the right age to answer that question eludes workers...." --*Scientific American* **13**(2)

Views of Old Earth/Progressive Creationists

- Consider Neanderthals, Homo erectus, and others a ________ creation
- Call them "_______ _________ to distinguish them from human
- Incorrectly claim no direct link between humans and ____________
- Typically accept evolutionist's dating

____

- Study of ______________

How are human fossils studied?

- Often "pieced" together
- Not studied ________, use of casts
- Fragile and irreplaceable

Donald Johanson

- "Only those in the ____________ get to see the fossils; only those who agree with the particular interpretation of a particular investigator are allowed to see the fossils."

Most investigators use casts

- May not be ________
  - piltdown man--filed teeth

Piltdown man hoax

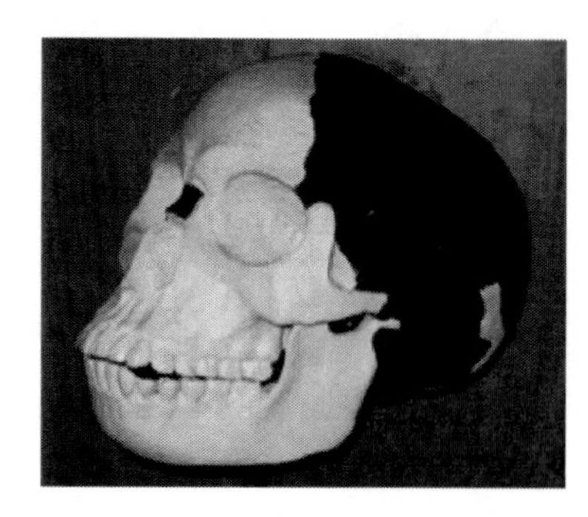

Piltdown

- _____________, human skull cap
- Ales Hidlicka-- jaw is nearer to humans than the apes
- Elliot Smith-- skull is primitive and had an ape-like brain

Common Portrayal of the Neanderthal

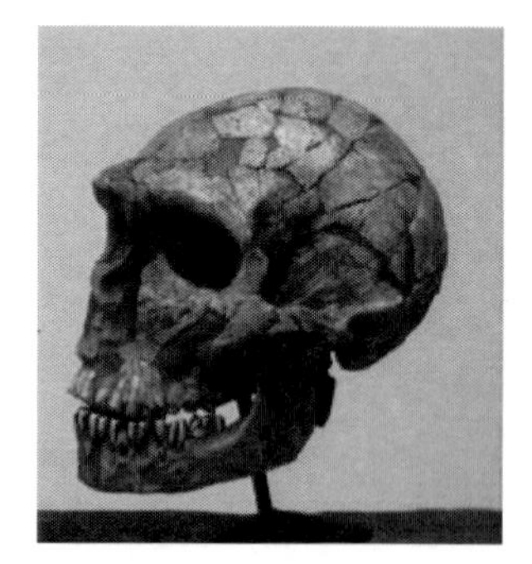

- Neanderthals usually viewed as:
  - Less __________
  - "brutish"
  - Typical cave man
  - VERY different from modern humans
- Probably ________

Sites of Neanderthal Inhabitation

- Remains have been found across _________, especially ______
- Some have been found in Croatia, Iraq and Israel

Distinct Neanderthal Characteristics

What Did the Neanderthals Do?

- Used ____ and tools (Knives, scrapers, etc.)
- ________ and trapped prey
- Used _______
- Probably _________ with "modern humans"
- Cared for each other
- ___________________

Neanderthal?

- Looks like somebody from the W.W.E.

How have evolutionists attempted to solve the problem?

- Fossils and archeological discoveries have ______ to solve the relationship of Neanderthals
- Therefore some scientists appeal to ___ evidence

Neanderthals and Humans ***DID*** Share a Common Gene Pool

- "The separate phylogenetic position of Neandertals is not supported when these factors are considered. Our analysis shows that the Neandertal-Human and Human-Human pairwise distance distributions overlap more than what previous studies have suggested."
  - Gutierrez et al. *Mol. Biol. Evol.* 19:1359 2002

Implications for Modern Human Origins and Neanderthal Genetics

- The divergence time between Neanderthals and modern humans is ________.
- Neanderthals ______ be excluded from human history.

Homo erectus "upright man"

- 0.3-1.8 m.y.a.
- 1891, Eugene ______, Java Island
- skull cap, femur, few teeth found
- 1920's admitted finding _____ 1 & 2
  - similar to modern man, same strata
- Java man _________ be ancestral to modern humans

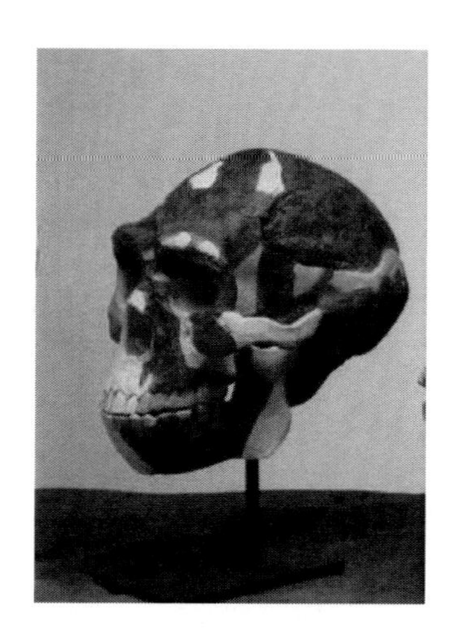

Java man

Homo erectus

- Previously thought 1.6 m.y.a to 400,000 however recent dating in Java indicates only _____ years (contemporary with modern man)
  - 222 homo erectus fossils found
  - 108 date 0.3-1.8 m.y.a.
  - 24 date 1.5-2 m.y.a.
  - 106 date < 300,000 y.a.
  - 62 date < 12,000 y.a.

Homo ergaster

- *Homo erectus* specimens found in ______ are assigned to this group
- Modern humans share features with *ergaster* which are _________ from *homo erectus*
- Assigned to 1.9-1.5 million years ago
- Assumed to be ancestral to Homo populations but likely just regional differences

Homo Habilis

- Some KNM-ER 1470 similar to homo sapiens (really more ape-like, upon closer inspection)
- Others (KMN-ER 1805) small and ____ like.
- Overlaps homo erectus therefore not a good ancestor
- No longer considered to be ancestral to any members of *Homo*

Homo rudolfensis

- KNM ER 1470 and others formerly classified as Homo habilis placed into a new category because of _________________ with other habilis specimens
- Status still a mater of debate
- Alleged ~2.4-1.8 m.y.a.

Australopithecus--Lucy

- Creationists and some evolutionists consider it an __________.
- Only __% of Lucy's bones were found
- What about the Kanapoi elbow?

Does this sound like good science?

- "The humeral fragment from Kanapoi, with a date of about 4.4 million could not be distinguished from *Homo sapiens* morphologically....We suggested that it might represent Australopithecus because at that time allocation to *Homo* seemed preposterous, <u>*although it would be the correct one without the __________.*</u>"

Lucy "reconstruction" from MSU

Look at the human hands and human feet!
No bones from the hands and feet of Lucy were found!

This is a reconstruction of the famous fossil hominid "Lucy", one of our oldest known ancestors. It is based on a fossilized and almost complete skeleton found in the Hadar region of Ethiopia by Dr. Donald Johanson. She has been dated to ca. 3.18 million years ago and is classified as *Australopithecus afarensis*. Lucy and other fossil and living primates and hominids will be the focus of an upcoming MSU Museum exhibition on Primate Evolution.

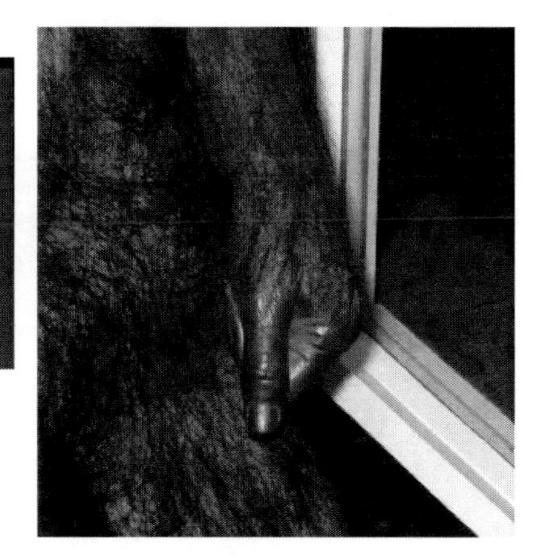

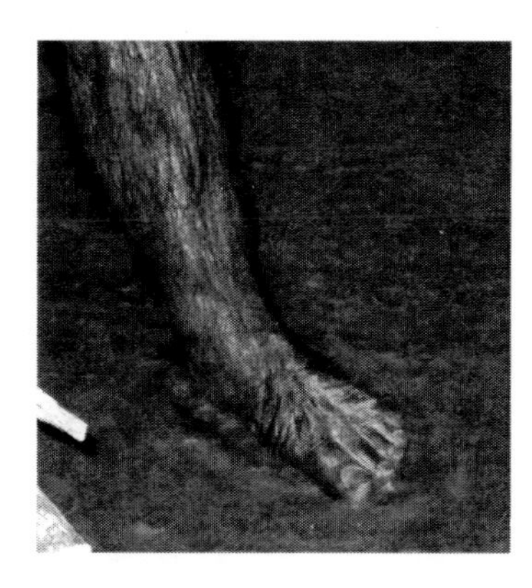

Do New Fossils Shed Light on Human Origins?

- "Newly discovered fossils from Chad, Kenya and Ethiopia may extend the human record back to seven million years ago, revealing the earliest hominids yet."
- "These finds cast doubt on conventional paleoanthropological wisdom. But experts disagree over how these creatures are related to humans—*if they are ____________l.*"

– **Scientific American** ***13:6 2003 (emphasis added)***

Evidence of human ancestor?

- Even ____ evidence for a common ancestor of humans and apes than anything else in evolution

"Consequently, just as the result of ___ trespass was condemnation for all men, so also the result of ___ act of righteousness was justification that brings life for all men. For just as through the disobedience of the ___ man the many were made sinners, so also through the obedience of the ___ man the many will be made righteous." Romans 5:18-19 (NIV)

Lesson 19

Lucy, She's No Lady

Notes

Notes

Lesson 20
Cellular Complexity

Watchmaker Argument

"If you find a _____ on the beach, you know that there was a __________. You know that it didn't arise spontaneously."

Design is not "God of the gaps"

Things which are not understood (gaps) are attributed to ___
But….
When we understand them, then God seems to be smaller or irrelevant to some.

'It is not that the methods and institutions of science somehow ______ us to accept a material explanation of the phenomenal world, but, on the contrary, that we are forced by our ________ adherence to material causes to create an apparatus of investigation and a set of concepts that produce ________ explanations, no matter how counter-intuitive, no matter how mystifying to the uninitiated. Moreover, that materialism is an absolute, for we ______ allow a Divine Foot in the door.'
Richard Lewontin, Billions and billions of demons, *The New York Review*, p. 31, 9 January 1997.

- "Biologists must constantly keep in mind that what they see is not ________, but rather evolved."
  – Francis Crick, *What Mad Pursuit* (New York: Basic Books, 1988), 138

___________ Complexity

- What is minimally required for function?
- Such a system could not arise from ____________ of "simpler" components
- _______ _________ cannot be used because no _________ for individual components.
- System must have been in place ____________.

Example: Mouse Trap

May be modified or improved (natural selection) but certain features are minimally ________ for function.
No selective _________ for intermediate/precursor, all components must be present and functional.

Bacterial _________ is Irreducibly Complex

Ken Miller Argues Otherwise…

- Suggests Type III Secretory System was a _________
- Instead of a _________, cell has a "microsyringe" to inject toxins into other cells
- At least __ proteins show strong similarity to proteins of the ____________
- Evolution of __________ resulted from modification of the Type III system

…But the __________ Came _____!

- "We suggest that the _________ apparatus was the evolutionary precursor of Type III protein secretion systems."

Phylogenetic analyses of the constituents of Type III protein secretion systems.
Nguyen L, Paulsen IT, Tchieu J, Hueck CJ, Saier MH Jr.
J Mol Microbiol Biotechnol. 2000 Apr;2(2):125-44

If there is a creator, why is there disease?

- Many diseases result from a ____ of genetic information
- An example of "__________"

Darwin vs. Irreducible Complexity

- "If it could be demonstrated that any complex organ existed which could not possibly have been formed by ________, successive, ______ modifications, my theory would absolutely ________________."

– *Charles Darwin,* **Origin of Species**

“numerous, successive, slight modifications”

- Almost anything can appear to follow an evolutionary progression.
- __________ does not “prove” common ________
- But where do they come from first?
  - Dependent or independent origin
- Where does the information come from?

“numerous, successive, slight modifications”

Proteins

- _______ chains of amino acids
- Specific sequence using __ different amino acids
- about _______ amino acids long

“Proteins are not just rigid lumps of material with chemically reactive surfaces. They have precisely engineered moving parts whose mechanical actions are coupled to chemical events.” Alberts et al., Molecular Biology of the Cell

Protein Structure

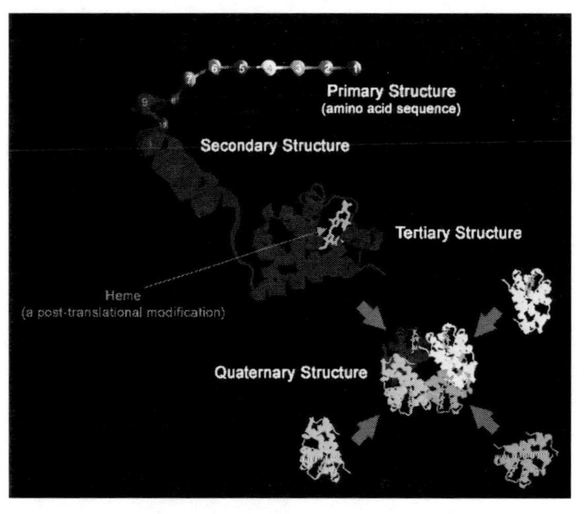

- The three dimensional ______ of a protein determines its _______
- The amino acid _______ determines the three dimensional _________
- The ___ sequence determines the amino acid sequence

- Proteins range 100-800 amino acids
- Average protein is ~___ amino acids
- Number of possible proteins ~ 100 AA long = $20^{100} = 10^{130}$
- Estimated number of hydrogen atoms in the universe ~ 10^{80}

Monkeys on Keyboards

- If there were enough monkeys typing on keyboards, given enough time, one of them would write a sonnet of __________

t Even with 14 billion years, some things just will ___ happen

Where does the _____________ come from?

- Recent report that scientists refuted creation claim
- Did not solve the problem
- Changing the number of ____ does not explain the origin of ____ and _____

Evolutionary Explanation for Source of New Genes

"The raw material of evolution is the DNA sequence that already exists: there is no natural mechanism for making long stretches of new random sequence. In this sense, no gene is ever entirely new." Alberts et al., MBoC 4th Ed.

However, most genes are "________" with no known origin.
"In bacillus subtillis, 2126 or __% of the genes have no family relationship or homology to anything else." Alberts et al., MBoC 4th Ed.

Metabolism

>___ different enzymes involved in metabolism

Biblical Concepts from Biology

"Go to the ant, you sluggard; consider its ways and be wise"
-Proverbs 6:6
Let's take this to the _________ level

Mitochondrial Permeability Transition Pore

When opened, allows cytochrome c (14 kD) protein to pass through
Opening this pore is an early event in some forms of apoptosis (or programmed cell death)

Potassium Channel

100 million K+ ions/second
Ions line up, single file
Selectivity filter only allows K+ at 1.33 A, but not Na+ at .95 A.
Very narrow pore, one atom at a time
"Enter through the narrow gate. For wide is the gate and broad is the road that leads to destruction, and many enter through it. But small is the gate and narrow the road that leads to life, and only a few find it."

-Matthew 7:13-14 NIV

Collagen triple helix

"A cord of three strands is not quickly broken"

Ecclesiastes 4:12 NIV

Laminin

Three stranded molecule
Shape of a cross
Present in the extracellular matrix—space *between* cells
Important for cell *adhesion* and *growth* of neuronal processes

"For there is one God and one mediator between God and men, the man Christ Jesus, who gave himself as a ransom for all men…"

1 Tim 2:5-6 NIV

God of the Gaps

- Now that we understand the details and the *complexity* of the details it points to the wisdom and design of the LORD!

Romans 1:20

- "For since the creation of the world God's invisible qualities--his eternal power and divine nature--have been clearly seen, being understood from what has been made, so that men are without excuse."

Notes

Lesson 21

Origin of Life

How did the first life arise?

- Answer from Evolution model:
  - Biological molecules, then cells arose from a ____________
- Answer from Creation model:
  - each living thing specially created, fully formed
  - made to reproduce after its kind

Spontaneous Generation

- Belief that living things could arise from _________ material
- Late 1800's _______ and _______ definitively showed that spontaneous generation was false.

The Cell Theory

- All living things are made of _____.
- The cell is the fundamental unit of ____.
- "*omnis celula e celula*"
- Every cell comes from a ____________ cell!

DNA to Proteins (via RNA)

- Central Dogma of Biology (Genes)
- ___ is ____________ into _______
- ________ is translated into ____________ sequence of proteins

DNA —Transcription (in the nucleus)→ RNA —Translation (in the cytoplasm)→ Protein

Hereditary Material

♦ Central Dogma
♦ Sequence of amino acids coded for in DNA/RNA
♦ ________ are required for making ___ and ___
♦ Too simple!

DNA Structure

t DNA is a linear polymer of ___________
t Double stranded, sugar phosphate backbone
t ____: stretch of DNA which codes for a protein
t Therefore, DNA serves as ____________ for making protein

DNA Must Be Replicated

t Found in the _______
t Semi-conservative
t DNA unwinds,
t DNA polymerase and other enzymes add free nucleotides, complimentary to template

RNA: Ribonucleic acid

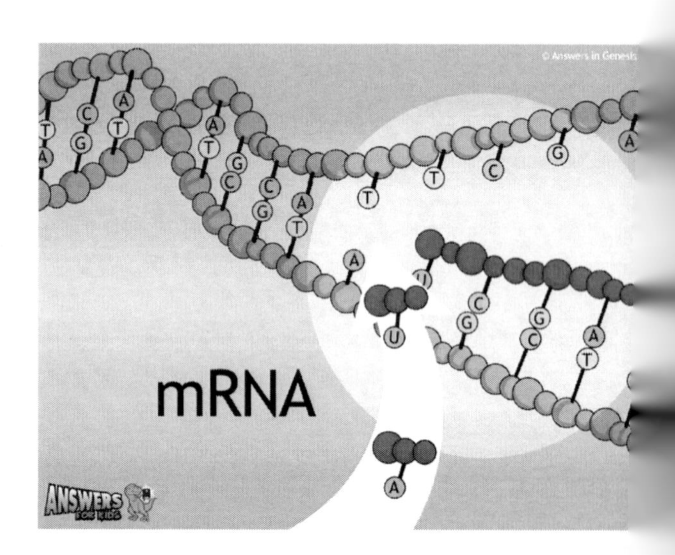

t _ types of RNA
  – ____: messenger RNA
  – ____: transfer RNA
  – ____: ribosomal RNA
t _________ stranded
t Ribose instead of deoxyribose
t A, G, C, _

RNA _______________

♦ RNA polymerase: >__ different proteins
♦ Must have RNA polymerase to use a DNA based system.

___________: Protein Synthesis

- Requires:
- ________ >50 different proteins
- mRNA with codons
- tRNA with anti-codon and amino acid

t-RNA

- Amino-acyl-tRNA synthetase

The Genetic Code

Making a protein requires

1. ____ instructions
2. mRNA intermediate
3. _________
   (protein factory)
4. tRNAs with amino acids

Mutations

- Change in stop/start codon
- Change in promoter, regulation
- Change in __________ sequence

But…

- DNA replication, RNA transcription, protein synthesis and amino acid synthesis require _______! (proteins)

- How do we get proteins without DNA?
  How do we get DNA without proteins?

Could DNA, RNA, or proteins arise ____________ to form a living cell?

Spontaneous Generation?

- ____ was the first biological molecule made in a ___ (1800's)
- Miller & Urey synthesized amino acids

Difficulties with the Miller experiments

- Reason for starting material? No ______
- Both L and D amino acids are formed
- Useless products also
- Trap protects amino acids
- Side reactions are favored

Other difficulties

- Making amino acids does not = proteins
- random proteins are _______ anyway
- Proteins must have a ________ amino acid sequence for function PRTOIEN
- Proteins are only made of ____ (L) handed amino acids
- Conclusion: Proteins could not form spontaneously

DNA could *not* form spontaneously either

- No mechanism to generate protein from ___ directly
- No mechanism for DNA replication without _______ (proteins)
- DNA impossible to make without________!
- Not DNA; not proteins; what's left?

RNA world?

- Proposes RNA first
  - Proteins are made from RNA

- – DNA can be made from RNA
- – RNA can have "enzymatic" activity
- t Reasons against spontaneous RNA
  - – RNA _____ form spontaneously
  - – ______ is required but not easily made
  - – No mechanism for sequence ___________

Problems with an RNA world

- t 1. RNA must do everything
- t 2. Reduced ability to ______
- t 3. Lack of ___________ to link amino acids
- t 4. How to convert to DNA system
- t 5. How to transfer function to ________

What's left?

- t What's left since proteins, DNA, and RNA ______ arise through natural processes?
- t Furthermore, a living organism requires more than DNA, RNA, and proteins to be alive.
- t Francis Crick suggests life came from ___________.

DNA Shows the Wisdom of God

- ♦ When we take a closer look at what God has made, it looks _____________
- ♦ DNA shows that we are truly "fearfully and wonderfully made"

Notes

Lecture 22
Creation Evangelism